Y0-ARK-571

Solid Phase Microextraction

Solid Phase Microextraction
Theory and Practice

Janusz Pawliszyn

New York • Chichester • Weinheim • Brisbane • Singapore • Toronto

Janusz Pawliszyn
Department of Chemistry
University of Waterloo
Waterloo, Ontario, Canada N2L 3G1

This text is printed on acid-free paper. ∞

Library of Congress Cataloging-in-Publication Data

Pawliszyn, Janusz.
Solid phase microextraction : theory and practice / Janusz Pawliszyn.
p. cm.
Includes bibliographical references (p. 229–242) and index.
ISBN 0-471-19034-9 (alk. paper)
1. Extraction (Chemistry) 2. Chemistry, Analytic—Technique.
I. Title.
QD63.E88P39 1997
543′.089—dc21 96-45276
CIP

Printed in the United States of America

ISBN 0-471-19034-9 Wiley-VCH, Inc.

10 9 8 7 6 5 4 3 2 1

To Barbara, Brian, Michael, Christina, Richard, members of my research group at the University of Waterloo, and my colleagues at Supelco, Varian, the Waterloo Center for Groundwater Research and the Department of Chemistry for help in the development of the SPME technology and this book.

Preface

Recently emphasis has been placed on developing solvent-free sample preparation methods because of new regulatory restrictions on the use of solvents, as well as the realization that a substantial saving of time and increased analytical performance can be achieved. New and established solvent-free techniques that use gas, membrane, or sorbent extracting phases are being investigated for various applications. The need to validate and optimize these new methods has also sparked research into the fundamentals of extraction processes. In-depth investigations are being conducted on the interactions between native analytes and sample matrices to find conditions under which the analytes are quantitatively released. The simplification of sample preparation and its integration with both sampling and convenient introduction of extracted components to analytical instruments is a significant challenge and an opportunity for the contemporary analytical chemist. The results of current research will have a profound effect on future analytical technology. This monograph describes one of the recently developed solvent-free sampling/sample preparation/introduction techniques: solid phase microextraction (SPME), winner of the 1994 R&D 100 Award.

Acknowledgments

The SPME work in my laboratory was supported by grants from the Natural Sciences and Engineering Research Council of Canada, Supelco, Varian, U.S. Environmental Protection Agency Office of Exploratory Research, Dow Chemical, Imperial Oil, the Waterloo Center for Groundwater Research, and the Ontario Ministry of Environment and Energy.

Janusz Pawliszyn
Ontario, Canada

Contents

CHAPTER

1

Solid Phase Microextraction in Perspective

1.1 Overview

The analytical procedure has several steps: sampling, sample preparation, separation, quantitation, statistical evaluation, and decision (see Figure 1.1). Each step is critical for obtaining correct results. The sampling step includes deciding where to get samples that properly define the object or problem being characterized, and choosing a method to obtain samples in the right amounts. For example, if a shipment of iron ore is received by a mill, a representative sample of an ore needs to be obtained from the materials transported before appropriate analysis can be done. On the other hand, the investigation of a toxic spill calls first for a decision to be made about the source of the sample, then the extent of the biological impact of the spill can be estimated based on analysis. A sample preparation step is necessary to isolate the components of interest from a sample matrix because most analytical instruments cannot handle the matrix directly. Sample preparation can include "clean up" procedures for very complex "dirty" samples. This step must also bring the analytes to a suitable concentration level. For example, before the determination of trace components present in soil or water can be determined, they must be isolated from the matrix, then concentrated, and frequently subjected to clean up.

During the separation step of the analytical process, the isolated complex mixture containing target analytes is divided into its constituents, typically by means of a chromatographic or electrophoretic technique. Quantitation is the determination of amounts of the identified compounds. The identification

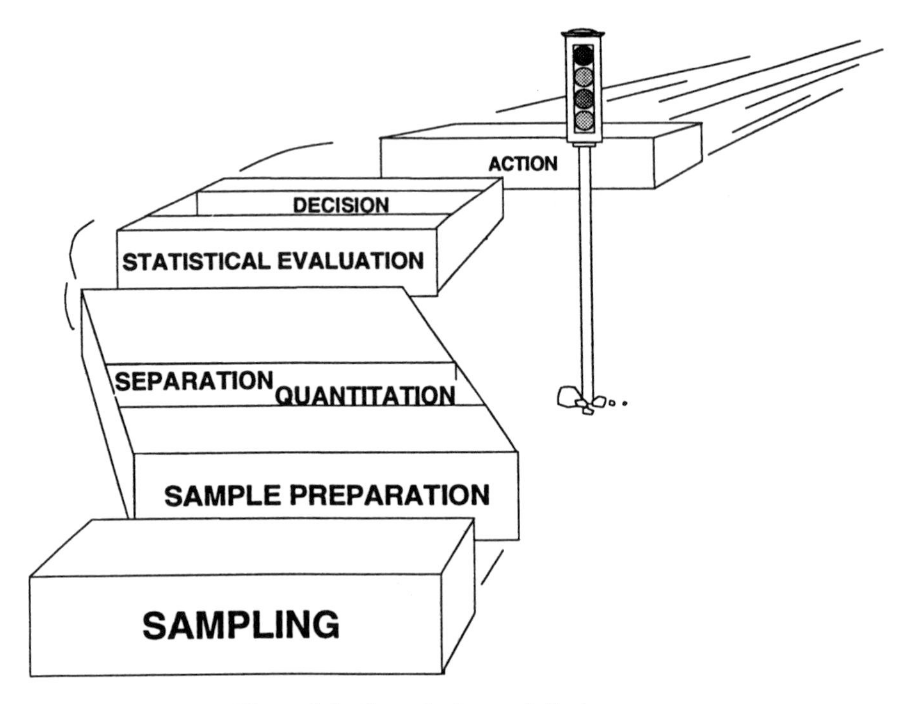

Figure 1.1 Steps in the analytical process.

can be based on a retention time combined with selective detection; more frequently, however, more specific instruments, namely, mass spectrometers, are used to eliminate possible errors in quantification due to interferences. Statistical evaluation of the results provides an estimate of the target compound concentration in the sample being analyzed. The data, will then indicate certain decisions, which might include a move to take another sample for further investigation of the object or problem.

It is important to note, as emphasized by Figure 1.1, that analytical steps follow one after another, and the next one cannot begin until the preceding one has been completed. Therefore, the slowest step determines the overall speed of the analytical process, and improving the speed of a single step may not result in a throughput increase. To increase the throughput of analysis, all steps need to be considered. Also, errors performed in any preceding step, including sampling, will result in the overall poor performance of the procedure.

An ideal instrument would perform all the analytical steps without human intervention, preferably directly in the field. Although such a device has not yet been built, today's sophisticated instruments, such as the gas chromatograph/mass spectrometer (GC/MS), can separate and quantify complex mixtures and automatically apply chemometric methods to statistically evaluate results. Despite the advances in techniques of separation and quantitation, many sampling and sample preparation practices are based on nineteenth century technologies, such as the common Soxhlet extraction method.[1] Traditional sample preparation methods are typically time and labor intensive, have multi-step procedures prone to loss of analytes, and use toxic organic solvents. These characteristics make such methods very difficult to integrate with sampling and separation methods for the purposes of hyphenation and automation. The result is that over 80% of analysis time is spent on sampling and sample preparation steps.

Substantial improvements in this area of instrumentation will translate immediately into important time-saving and convenience enhancements. Also, a new awareness of the hazards of solvents—the risk of cancer and the depletion of ozone—has resulted in international initiatives such as the Montreal Protocol to stop the production of many organic solvents. If for no other reason, the phasing out of solvents is poised to induce a major change in analytical methodology.[2] This change is an opportunity for the scientific community to correct the problems of the current sample preparation methods and to formulate new and practical solvent-free alternatives.[3]

Any sample preparation method, of course, should have good analytical performance characteristics, including efficiency, selectivity, and applicability to various compounds and matrices. A new method should allow simple automation and field analysis. An ideal sample preparation technique would be easy to use, inexpensive, and compatible with a range of analytical instruments, since the costs of training and new hardware frequently prevent analytical practitioners from adopting new methods. This monograph is dedicated to a relatively new solvent-free sample preparation method, solid phase microextraction (SPME).

1.2 Solvent-Free Sample Preparation Techniques

The operating principle of any sample preparation method is to partition analytes between the sample matrix and an extracting phase. Sample preparation techniques that use little or no organic solvent have been available for some time. They can be classified, as shown in Figure 1.2, according to the extracting phases of gas, membrane, and sorbent.

Gas Phase Sample Preparation Methods. Gas phase sample preparation methods, such as static headspace sampling or purge-and-trap extraction, have as a common feature, the partitioning of analytes into a gas phase. During the partitioning, nonvolatile high molecular weight compounds are eliminated, which prevents contamination of the separation column and therefore makes this method very rugged. Headspace sampling has been widely used to analyze volatile compounds because the extracting phase (air, helium or nitrogen) is compatible with most instruments, such as gas chromatographs. In the static headspace procedure, a sample is simply allowed to equilibrate with its heads-

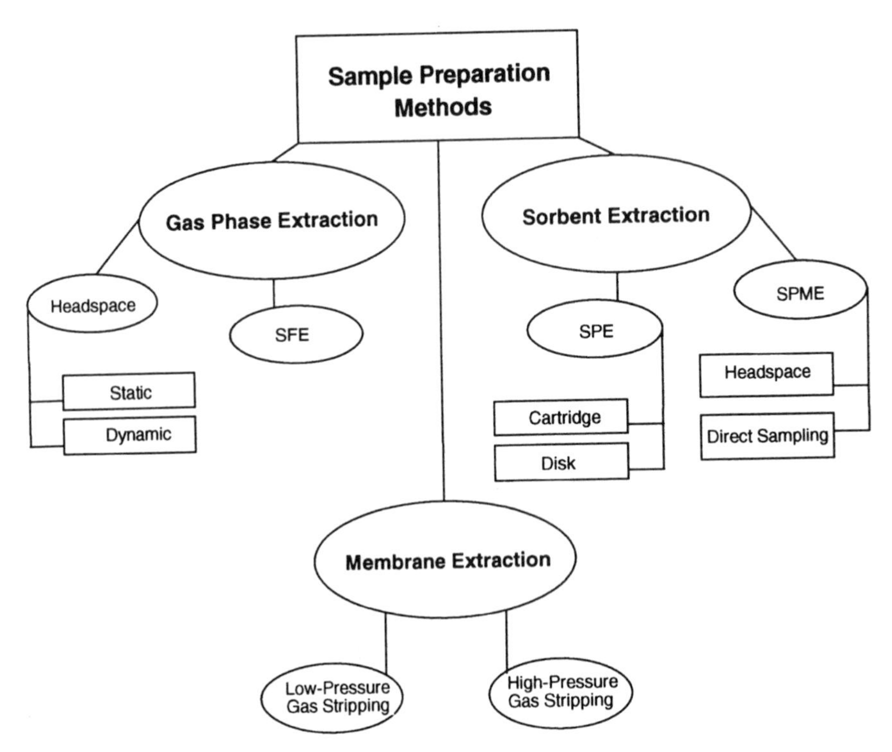

Figure 1.2 Classification of solvent-free sample preparation methods. Reprinted with permission from J. Pawliszyn, *TrAC* **14**, 113 (1995).

pace and then a small, well defined volume of the headspace is directly injected into a GC for analysis. Static headspace sampling is probably the simplest and, to date, the most frequently applied solvent-free sample preparation technique, particularly in field analysis. The amount of analyte transferred to the instrument is proportional to the volume of the gaseous phase, the Henry constant, and the concentration of analyte, assuming the headspace to be in equilibrium with the sample.

Static headspace methods have been used to analyse volatile organic compounds (VOCs) in food, beverage, clinical, and other samples.[4,5] The static headspace method can be applied to "sniff" samples in the field without placing them in containers. Simple instruments that "sniff" air directly above an investigated area are based on static headspace. The technique is low in sensitivity, however, because it lacks a concentrating effect. Static headspace cannot achieve exhaustive extraction, except in the case of very volatile gases, and therefore requires careful calibration. Dynamic headspace uses multiple processes and allows quantitative removal of VOCs. The commonly used purge-and-trap approach, a dynamic headspace method for the analysis of VOCs in water, has two steps.[6,7] The first step is to let a carrier gas purge through an aqueous sample to remove VOCs from the matrix. The second step is to quantitatively collect these compounds by using a cold or a sorbent trap. This technique has some drawbacks including foaming, carryover of analytes from a previous determination, and the fact that the stripping flow rate is incompatible with the separation instrument.

In combination with thermal desorption, the headspace approach can be used for analysis of less volatile compounds. When the sample is heated to an elevated temperature, analytes are thermally desorbed from the matrix and more efficiently partitioned into the gas phase. Heating is particularly important in the analysis of samples containing solids, such as clay soil, which tend to strongly bind organic analytes. However, thermally unstable analytes and a high moisture content in the desorbed gas mixture frequently prevent the use of the thermal desorption approach.

Another gas-based sample preparation method uses liquid such as compressed carbon dioxide as an extracting phase that is capable of removing less volatile compounds at ambient temperatures. This technique, called supercritical fluid extraction (SFE), has been rapidly developed and has some attractive characteristics because supercritical fluids possess both gas like mass transfer and liquid like solvating characteristics.[8,9,10] However, SFE requires an expensive high-pressure fluid delivery system and high purity gas source, both of which are heavy equipment that make field analysis difficult, at present. Since this technique can extract nonvolatile compounds at ambient temperatures, it is useful for the analysis of thermally unstable analytes and matrices, particularly foodstuffs.

Membrane Extractions. Membrane extraction consists of two simultaneous processes: extraction of analytes from the sample matrix by the membrane material, and extraction of analytes from the membrane by a stripping phase. This method has been developed for mass spectrometry over the last three decades, but little information is available on its use for sample preparation for chromatography. In the past, nitrogen stripping gas was used to transfer the permeated analytes from the surface of a flat polymeric membrane to a bed of activated charcoal.[11] The compounds were desorbed into a GC for analysis after switching of a valve,. Although many early methods used supported membrane sheets, most recent developments of membrane extraction techniques have focused on the use of hollow fibers.[12,13,14] Hollow fiber membrane modules are simpler to make because a hollow fiber is self-supporting. Compared to membrane sheets and headspace methods, hollow fibers provides a higher ratio of surface area to volume for the stripping gas, which allows a more efficient extraction. Membrane extraction can be directly combined with MS or GC to perform continuous monitoring.[13,15]

A direct interface method, membrane extraction with a sorbent interface (MESI) system, consists of a hollow fiber membrane module, a low heat capacity sorbent interface, a separation instrument such as a capillary GC, and a computer (Figure 1.3).[13] The membrane is in direct contact with a sample

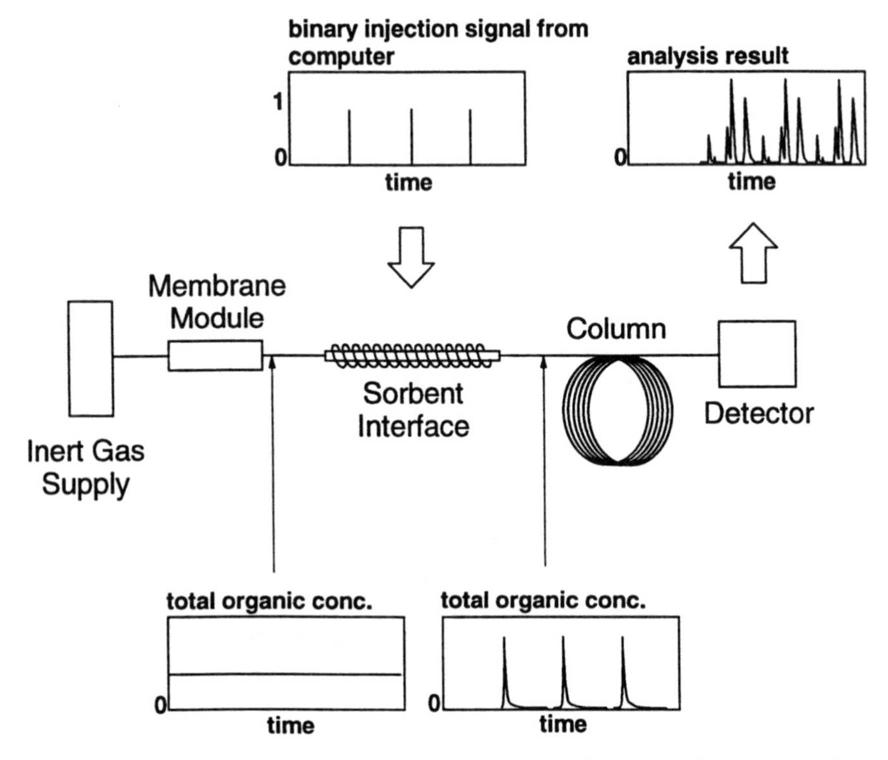

Figure 1.3 Components of the MESI system coupled to a gas chromatograph.

or its headspace. Using a hydrophobic membrane prevents moisture from entering the carrier gas. Analytes are concentrated on the surface of the membrane, transported across the membrane through diffusion, are stripped and carried by the gas to the trap. A heat pulse desorbs analytes collected at the trap and produces a narrow analyte band at the front of the separation column. When a low heat capacity sorbent, such as a piece of GC capillary column is used, the resulting injection profile is sharp, the analysis time is short, and close to real-time monitoring is achievable.[16] The direct sample-to-instrument route of the MESI system prevents loss of analytes.

Various designs of membrane modules can be used to analyze air, water, or soil headspace. The MESI approach is particularly suitable for continuous monitoring and process control. The membrane can be fitted conveniently into a flowing stream. The transport of analytes through the membrane introduces selectivity to the sample preparation process and, as in the headspace method, the membrane protects the separation column from high molecular weight compounds, with an additional sorbent concentration advantage. Membrane extraction is not limited to volatile compound analysis; indeed, extraction of higher molecular weight compounds can be performed by using higher temperatures or microporous membranes with various pore diameters. The membrane method has been applied to the analysis of semivolatile compounds by means of a high pressure stripping gas.[17]

Sorbent Extractions. The concept of using an adsorbent material to extract trace organic compounds from an aqueous sample was developed in the 1980s, and its application has been extensively reviewed.[18,19] Sorbents are now used to extract organic compounds from various matrices including water, air, and even soil. A sorbent with a strong affinity towards organic compounds will retain and concentrate those compounds from a very diluted aqueous or gaseous sample. Many sorbents are specifically suited for the extraction of different groups of organic compounds with various degrees of selectivity.

One widely used sorbent technique is solid phase extraction (SPE). The first step of SPE is to pass a liquid matrix through a plastic cartridge (tube) or flat membrane (disk) containing sorbent dispersed on a particulate support to extract analytes together with interfering compounds.[20] Usually, a selective solvent is used to remove interferences first, and then another solvent is chosen to wash out target analytes. SPE has a number of attractive features compared to traditional solvent extraction. SPE is simple, inexpensive, and uses relatively little solvent. The interaction between sample matrix and analytes often results in low recovery, however, and solid and oily components in a sample matrix may plug the SPE cartridge or block pores in the sorbent causing it to become overloaded. Sorbents suffer from high carryover values, and batch-to-batch variation of the sorbents leads to poor reproducibility. The SPE technique is limited to semivolatile compounds with boiling temperatures substantially above that of the desorption solvent temperature.

The limitations of SPE can be overcome by dispersing a minute quantity of the extracting phase on a fine rod made of fused silica or other appropriate material. Earlier uses of a small amounts of the liquid phase in microextraction techniques had demonstrated improved performance over the large-volume approach. Although full removal of target analytes from the sample matrix is not obtained, the high concentrating ability and selectivity of the technique allows direct, high sensitivity analysis of the extracted mixtures.[21,22] The difficulty in handling small volumes of solvents comprises a major limitation of this approach. Application of sorbing material permanently attached to the fiber addresses this limitation and allows reuse of the same extraction phase. The resulting technique is called Solid Phase Microextraction (SPME).[23,24] The miniature cylindrical geometry of this setup allows rapid mass transfer during extraction and desorption, and prevents plugging.

The SPME process has two steps: partitioning of analytes between the coating and the sample matrix, followed by desorption of concentrated extracts into an analytical instrument. In the first step, the coated fiber is exposed to the sample or its headspace, which causes the target analytes to partition from the sample matrix into the coating. The fiber bearing concentrated analytes is then transferred to an instrument for desorption, whereupon separation

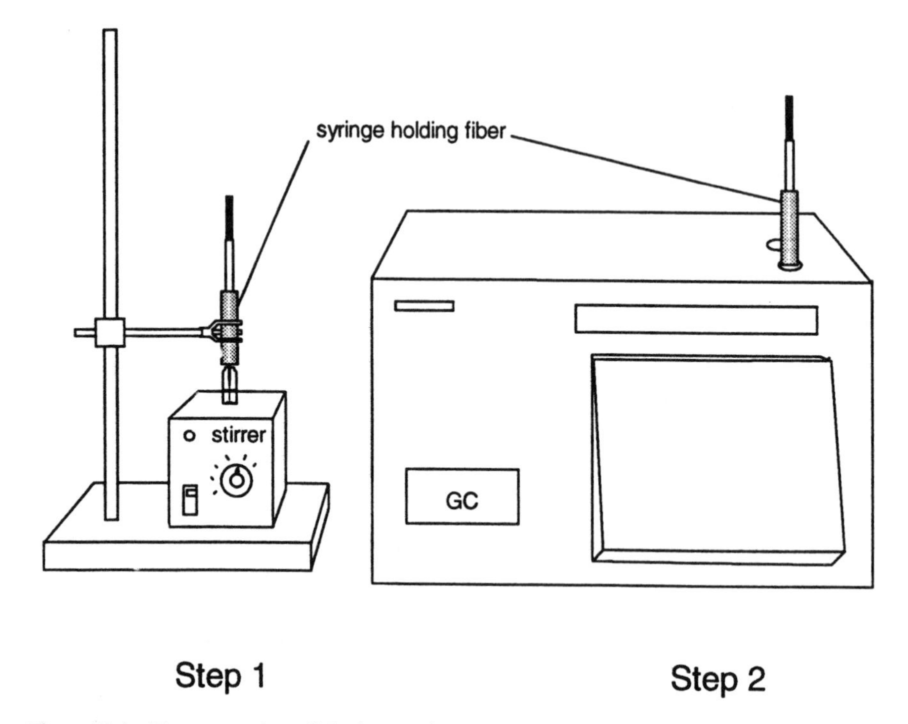

Figure 1.4 Two steps in solid phase microextraction: 1, equilibration of the analyte between the fiber coating and a sample matrix; 2, transfer of the fiber to an analytical instrument.

and quantitation of extracts can take place. A clean-up step using selective solvents could be incorporated, similar to SPE, but to date, this additional step has not been necessary, primarily because of the selective nature of the coatings. Figure 1.4 illustrates an SPME laboratory setup for direct extraction and GC analysis.

References

1. F. Soxhlet, *Dingers Polytech. J.* ***232***, 461 (1879).
2. D. Noble, *Anal. Chem.* ***65***, 693A (1993).
3. J. Pawliszyn, in R. Clement, K. Siu, and H. Hill (Editors), *Instrumentation for Trace Organic Monitoring*, Lewis Publishers, Chelsea, MI, 1992, Ch. 13, p. 253.
4. R.E. Kepner, H. Maarse, and J. Strating, *Anal. Chem.* ***36***, 77 (1964).
5. G. Charalambous, *Analysis of Food and Beverages, Headspace Technique,* Academic Press: New York, 1978.
6. K. Grob and F. J. Zucher, *J. Chromatogr.* ***117***, 285 (1976).
7. U.S. EPA Method 624, *Fed. Regist.* ***49***, 141 (1984).

8. S. Hawthorne, *Anal. Chem.* ***62***, 633A (1990).
9. J. Pawliszyn, *J. Chromatogr. Sci.* ***31***, 31 (1993).
10. T.L. Chester, J.D. Pinkston, and D.E. Raynie *Anal. Chem.* ***64***, 153R (1992).
11. R.D. Blanchard and J.K. Hardy, *Anal. Chem.* ***56***, 1621 (1984).
12. K.F. Pratt and J. Pawliszyn, *Anal. Chem.* ***64***, 2101 (1992).
13. M.J. Yang and J. Pawliszyn, *Anal. Chem.* ***65***, 1758 (1993).
14. M.J. Yang, S. Harms, Y. Luo, and J. Pawliszyn, *Anal. Chem.* ***66***, 1339 (1994).
15. T. Kotiaho, F.R. Lauritsen, T.K. Choudhury, R.G. Cooks, and G.T. Tsao, *Anal. Chem.* ***63,*** 875A (1991).
16. M.J. Yang and J. Pawliszyn, *LC-GC* ***14***, 364 (1996).
17. M.J. Yang and J. Pawliszyn, *Anal. Chem.* ***65***, 2538 (1993).
18. M. Dressler, *J. Chromatogr.* ***165***, 167 (1979).
19. C.F. Poole and S.A. Schuette, *J. High Resolut. Chromatogr.* ***6***, 526 (1983).
20. D. Hagen, C. Markell, G. Schmitt, and D. Blevins, *Anal. Chim. Acta* ***236***, 157 (1990).
21. D.A.J. Murray, *Chromatographia.* ***177***, 135 (1979).
22. D. R. Thielen, G. Olsen, A. Davis, E. Bajor, J. Stefanowski, and J. Chodkowski, *J. Chromatogr. Sci.* ***25***, 12 (1987).
23. C.L. Arthur and J. Pawliszyn, *Anal. Chem.* ***62***, 2145 (1990).
24. J. Pawliszyn, *Method and Device for Solid Phase Microextraction and Desorption,* PCT, International Patent Publication WO 91/15745 and national counterparts.

CHAPTER

2

Operating Principles and Construction of SPME Devices

The main objective of this chapter is to introduce the reader to the SPME techniques, devices and interfaces to analytical instrumentation that have been designed and evaluated to date. Chapter 3 provides a more detailed theoretical discussion of their operation. First-generation SPME commercial devices and coatings are briefly presented in Section 2.8.

2.1 Historical Perspective on the Early Work

Solid phase microextraction was developed to address the need to facilitate rapid sample preparation. The initial work on laser desorption/gas chromatography resulted in rapid separation times, even for very high molecular weight species.[1] However, the preparation of samples for this experiment took hours, which was over an order of magnitude longer than the separation times. In this experiment, optical fibers were used to transmit laser light energy to the GC instrument. The sample preparation process was analogous to standard solvent extraction procedures. The fiber tip was coated with the sample by dipping one end of the optical fiber in the solvent extract, coating the fiber and then removing volatile solvents through evaporation. The fiber tip, prepared in such a way, was inserted into the injector of a gas chromatograph, and analytes were volatilized onto the front of the GC column by means of a laser pulse. During that work, a need was recognized for rapid sample preparation techniques to retain the time efficiency advantages made possible by the use of laser pulse and a high speed separation instrument. The challenge was ad-

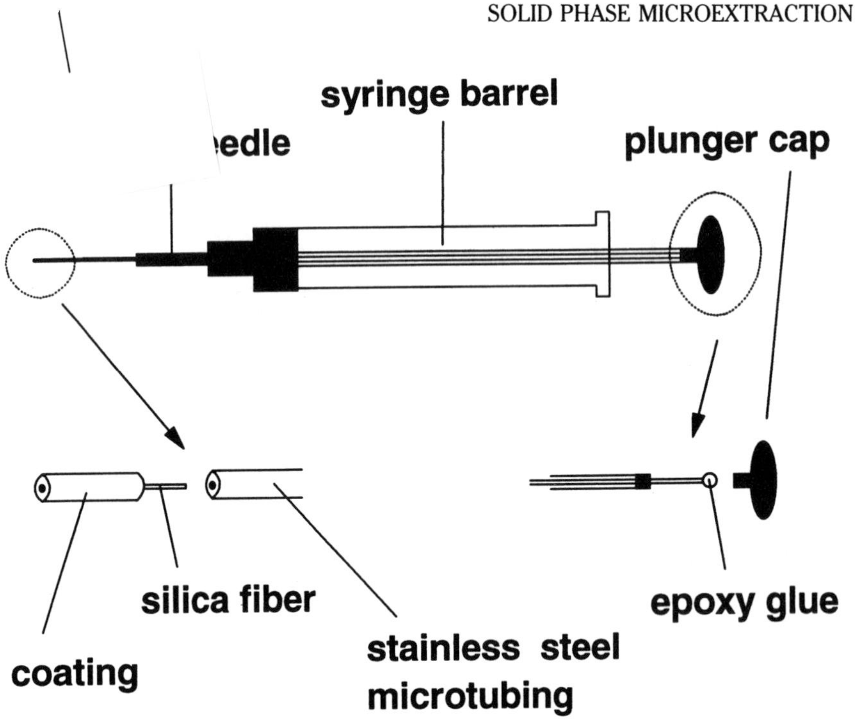

Figure 2.1 The custom-made SPME device based on Hamilton 7000 series syringe.

dressed using fibers, since optical fibers could be purchased coated with several types of polymeric film. The original purpose of these coatings was simply to protect the fibers from breakage. Because of the thin films used (10–100 μm), the expected extraction times for these systems were very short. In addition, novel types of film could be prepared, since chromatographers have a good knowledge base about fused silica coating methods gained from capillary columns manufacturing experience.

The initial work to test this concept was published in the spring of 1990. Sections of fused silica optical fibers, both uncoated and coated with liquid and solid polymeric phases, were dipped into an aqueous sample containing test analytes and then placed in a GC injector.[2] The process of introducing and removing the fibers required the opening of the injector which resulted in loss of head pressure at the column. Despite their basic nature, those early experiments provided very important preliminary data that confirmed the usefulness of this simple approach, since both polar and nonpolar chemical species were extracted rapidly and reproducibly, from aqueous samples.

The development of the technique accelerated rapidly with the implementation of coated fibers incorporated into a microsyringe, resulting in the first SPME device.[3] Figure 2.1 shows an example of the SPME device based on the Hamilton™ 7000 series microsyringe. The metal rod, which serves as the

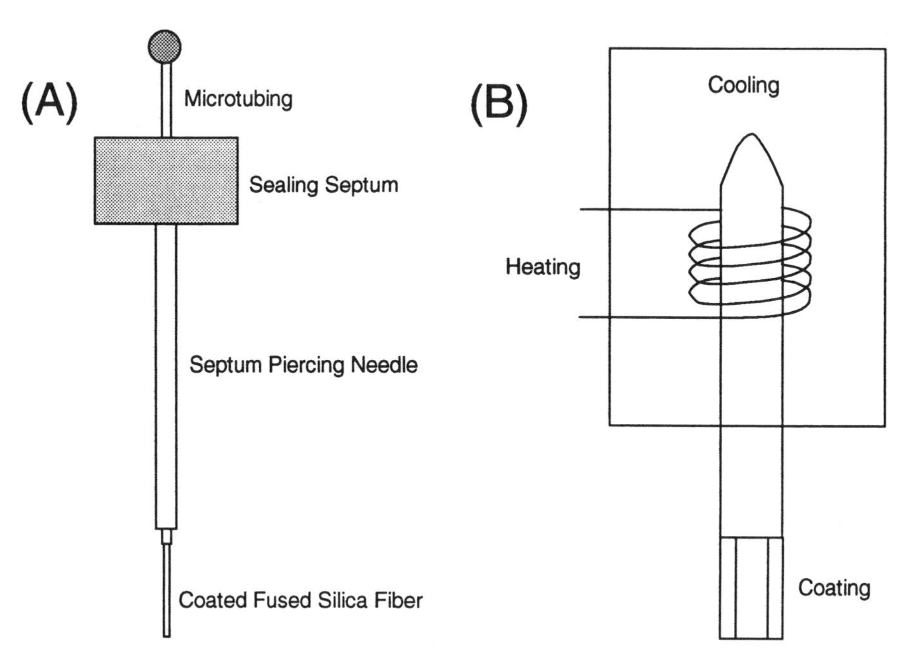

Figure 2.2 Simple versions of the SPME device using coated fibre (A) and internally coated tubing (B).

piston in a microsyringe, is replaced with stainless steel microtubing having an inside diameter (i.d.) slightly larger than the outside diameter (o.d.) of the fused silica rod. Typically, the first 5 mm of the coating is removed from a 1.5 cm long fiber which is then inserted into the microtubing. High temperature epoxy glue is used to permanently mount the fiber. Sample injection is then very much like standard syringe injection. Movement of the plunger allowed exposure of the fiber during extraction and desorption and its protection in the needle during stogage and penetration of the septa. SPME devices do not need expensive syringes like the Hamilton™ syringes. As Figure 2.2A illustrates, a useful device can be built from a short piece of stainless steel microtubing (to hold the fiber), another piece of larger tubing (to work as a "needle"), and a septum (to seal the connection between the microtubing and the "needle"). The design from Figure 2.2A is a basic building block of a commercial SPME device described later and illustated on Figure 2.25. Another simple SPME construction is based on a piece of internally coated tubing.[4] This tubing can be mounted inside a needle or it can constitute the "needle" itself of a syringe.[5] Elimination of mechanical movement of a plunger of a syringe can be accomplished by sealing the tubing at one end and installing a microheater as illustrated in Figure 2.2B. Expansion of air caused by temperature increase allows removal of desorbed analytes from the extracting phase located inside the tubing. An in-tube approach is useful in the design of passive

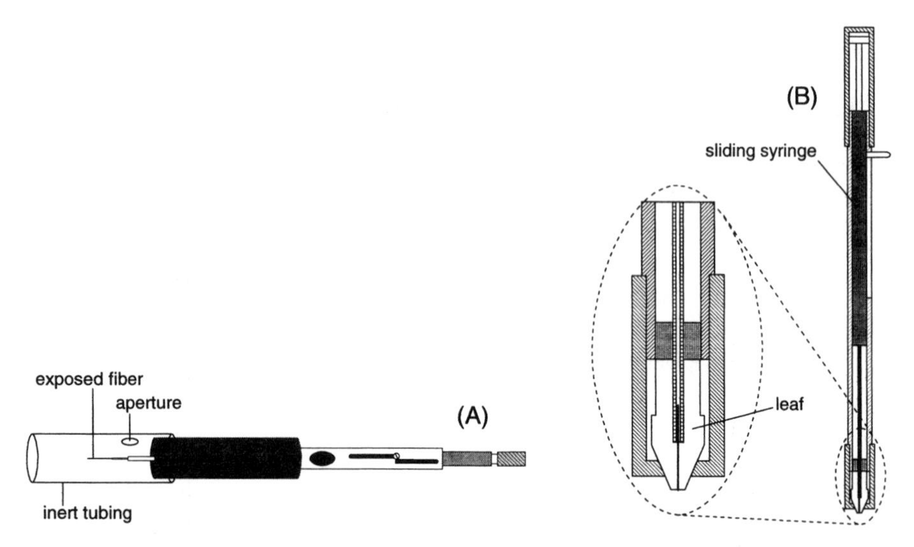

Figure 2.3 SPME device modified for breath analysis (A) and for the field application (B).

sampling devices discussed in Chapter 5, since, in this case, the extraction rate is limited by the diffusion of analytes into the needle.[6] In addition, active sampling is possible by heating and cooling the air contained in the upper part of the tubing. This causes movement of liquid or gaseaous samples in and out of the tubing, facilitating mass transport of analytes from the sample to the coating.

Although to date SPME devices have been used principally in laboratory applications, more current research has been directed toward remote monitoring, particularly for clinical, field environmental, and industrial hygiene applications. In their operating principles, such devices are analogous to the devices described above, but modifications are made for greater convenience in given applications. For example, as shown in Figure 2.3A, adding a tube with a small opening to cover the needle of the SPME syringe results in a useful device for breath analysis in a noninvasive clinical application.[7] This design can be further improved by adding a one-way valve mounted on the aperture, but the concept of operation remains the same.

An important feature of a field device is the ability to preserve extracted analytes in the coating. The simplest practical way to accomplish this goal is to seal the end of the needle with a piece of septum. Additionally, cooling extends the storage time. Polymeric septum material might cause the loss of analytes from the fiber, however. Therefore, a more appropriate approach is to use metal to metal seals. Figure 2.3B illustrates an example of construction based on "leafs" structure. It is anticipated that future devices designed for field applications will be more rugged than the current laboratory versions and will look more like "sticks" or "pens" than syringes.

2.2 Basic Principles of Solid Phase Microextraction

The transport of analytes from the matrix into the coating begins as soon as the coated fiber has been placed in contact with the sample (Figure 2.4). Typically, SPME extraction is considered to be complete when the analyte concentraton has reached distribution equilibrium between the sample matrix and the fiber coating. In practice, this means that once equilibrium is reached, the extracted amount is constant within the limits of experimental error and it is independent of further increase of extraction time. The equilibrium conditions can be described as:[8]

$$n = \frac{K_{fs}V_fV_sC_0}{K_{fs}V_f + V_s} \tag{2.1}$$

where n is the amount extracted by the coating, K_{fs} is a fiber coating/sample matrix distribution constant, V_f is the fiber coating volume, V_s is the sample volume, and C_0 is the initial concentration of a given analyte in the sample.

Equation 2.1, which assumes that the sample matrix can be represented as a single homogeneous phase and that no headspace is present in the system, can be modified to account for the existence of other components in the matrix by considering the volume of the individual phases and the appropriate distribution constants, as discussed in Chapter 3. The extraction can be interrupted and the fiber analyzed prior to equilibrium. To obtain reproducible data, however, constant convection conditions and careful timing of the extraction are necessary.

Simplicity and convenience of operation make SPME a superior alternative to more established techniques in a number of applications. In some cases, the technique facilitates unique investigations. Equation 2.1 indicates that there is a direct proportional relationship between sample concentration and

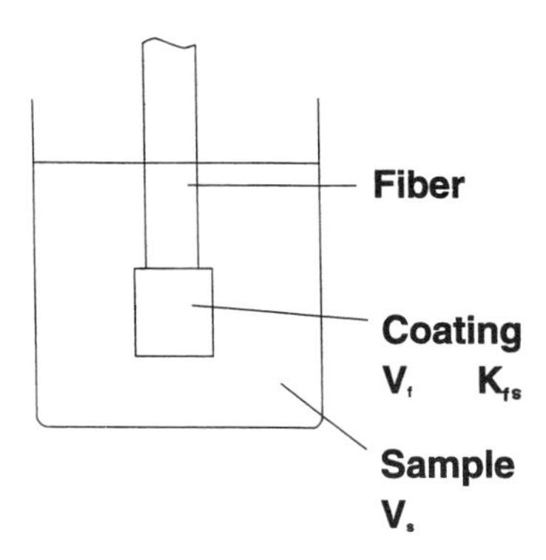

Figure 2.4 Microextraction with SPME.

the amount of analyte extracted. This is the basis for analyte quantitation. The most dramatic advantages of SPME exist at the extremes of sample volumes. Because the setup is small and convenient, coated fibers can be used to extract analytes from very small samples. For example, SPME devices are used to probe for substances emitted by a single flower bulb during its life span; the use of sub-micrometer diameter fibers permits the investigation of single cells. Since SPME is an equilibrium technique and therefore does not extract target analytes exhaustively, its presence in a living system should not result in significant disturbance. In addition, the technique facilitates speciation in natural systems, since the presence of a minute fiber, which removes small amounts of analyte, is not likely to disturb chemical equilibria in the system. It should be noticed, however, that the fraction of analyte extracted increases as the ratio of coating volume to coating increases. Complete extraction can be achieved for small sample volumes when distribution constants are reasonably high. This observation can be used to advantage if exhaustive extraction is required. It is very difficult to work with small sample volumes by using conventional sample preparation techniques. SPME allows rapid extraction and transfer to analytical instrument. These features results in additional adventage, when investigating intermediates in the system. For example, SPME was used to study biodegradation pathways of industrial contaminants.[9]

In addition, when sample volume is very large ($K_{fs}V_f \ll V_s$), equation 2.1 can be simplified to:

$$n = K_{fs}V_fC_0 \tag{2.2}$$

which points to the usefulness of the technique for field application. In this equation, the amount of extracted analyte is independent of the volume of the sample. In practice, there is no need to collect a defined sample prior to analyses as the fiber can be exposed directly to the ambient air, water, production stream, etc. The amount of extracted analyte will correspond directly to its concentration in the matrix, without being dependent on the sample volume. When the sampling step is eliminated, the whole analytical process can be accelèrated, and errors associated with analyte losses through decomposition or adsorption on the sampling container walls will be prevented. This advantage of SPME still needs to be explored practically, by developing portable field devices on a commercial scale.

2.3 Extraction Modes

Three basic types of extractions can be performed using SPME: direct extraction, headspace configuration, and a membrane protection approach. Figure 2.5 illustrates the differences among these modes. In the direct extraction mode (Figure 2.5a), the coated fiber is inserted directly into the sample and the analytes are transported directly from the sample matrix to the extracting

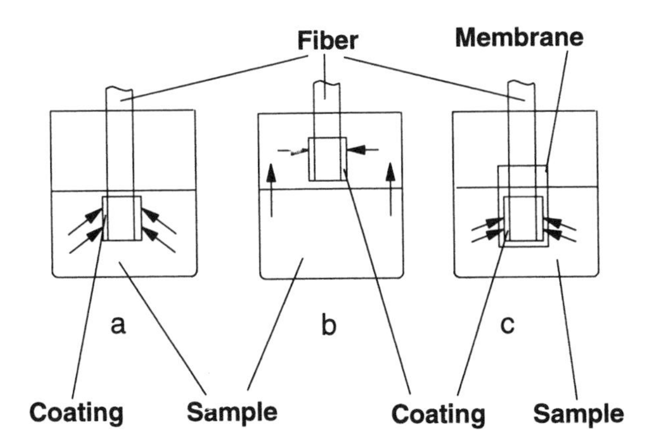

Figure 2.5 Modes of SPME operation: (a) direct extraction, (b) headspace SPME, (c) membrane-protected SPME.

phase. To facilitate rapid extraction, some level of agitation is required to transport analytes from the bulk of the solution to the vicinity of the fiber. For gaseous samples, natural convection of air is sufficient to facilitate rapid equilibration. For aqueous matrices, more efficient agitation techniques, such as fast sample flow, rapid fiber or vial movement, stirring or sonication are required.[6,10] These conditions are necessary to reduce the effect caused by the "depletion zone" produced close to the fiber as a result of fluid shielding and slow diffusion coefficients of analytes in liquid matrices. Chapter 3 discusses the effects of agitation in detail.

In the headspace mode, the analytes need to be transported through the barrier of air before they can reach the coating. This modification serves primarily to protect fiber coating from damage by high molecular weight and other nonvolatile interferences present in the sample matrix, such as humic materials or proteins. This headspace mode also allows modification of the matrix, such as a change of the pH, without damaging the fiber. Amounts of analyte extracted into the coating from the same vial at equilibrium using direct and headspace sampling are identical as long as sample and gaseous headspace volumes are the same. This is caused by the fact that the equilibrium concentration is independent of fiber location in the sample/headspace system. If the above conditon is not satisfied, a significant sensitivity difference between the direct and headspace approaches exists only for very volatile analytes.

The choice of sampling mode has a very significant impact on extraction kinetics, however. When the fiber coating is in the headspace, the analytes are removed from the headspace first, followed by indirect extraction from the matrix as shown in Figure 2.5b. Therefore, volatile analytes are extracted faster than semivolatiles since they are at a higher concentration in the headspace which contributes to faster mass transport rates through the headspace. Temperature has a significant effect on the kinetics of the process by determin-

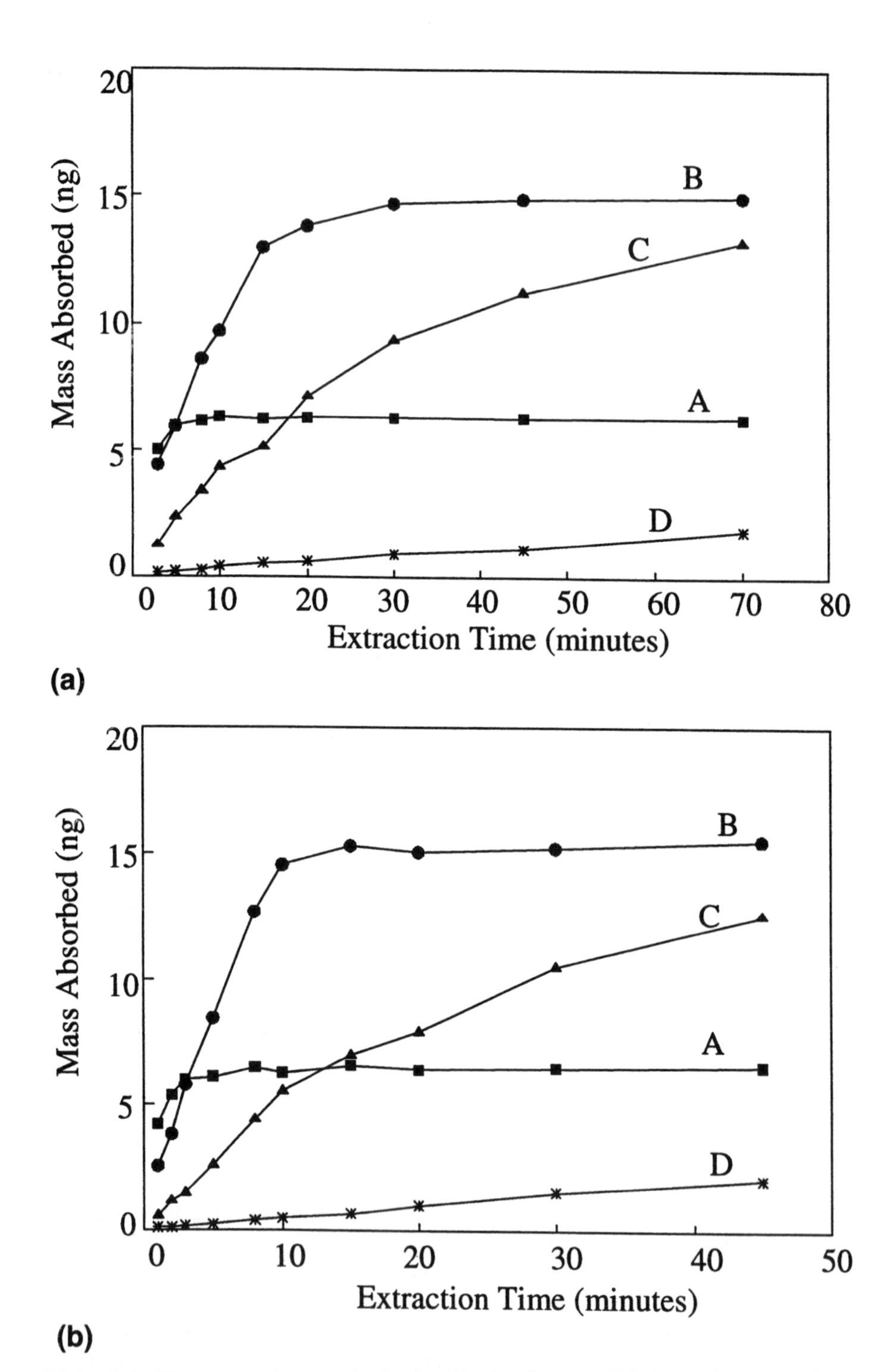

Figure 2.6 Time extraction profile obtained for headspace solid phase microextraction of several PAHs from aqueous samples at (a) 75% and (b) 100% stirring rates: A, naphthalene; B, acenaphthene; C, phenanthrene; D, chrysene.

Source: Adopted with permission from ref. 11.

ing the vapor pressure of analytes. In fact, the equilibration times for volatiles are shorter in the headspace SPME mode than for direct extraction under similar agitation conditions. This outcome is produced by two factors: a substantial portion of analytes is in the headspace prior to extraction, and that diffusion coefficients in the gaseous phase typically is four orders of magnitude larger than is in the liquid media. Since concentration of semivolatiles in the gaseous phase at room temperature is small, however, mass transfer rates, are substantially lower and result in longer extraction times. They can be improved by using very efficient agitation or by increasing the extraction temperature.[11]

Figures 2.6 a and b illustrate the effect of agitation on the extraction time profile obtained for polynuclear aromatic hydrocarbons (PAHs). As the rotational speed of the magnetic stirrer increases, the equilibration time of naphthalene and acenaphthene decreases from 8 minutes to 3 minutes and from 25 minutes to 10 minutes, respectively (comparing Figures 2.6a and 2.6b). For less volatile analytes, phenanthrene and chrysene, the equilibration is not reached during the experimental period in either case, but the amount of analyte extracted is about the same after 70 minutes extraction at low agitation (Figure 2.6a) as after 45 minutes of more efficient stirring (Figure 2.6b). Use of even more efficient agitation techniques, such as sonication, further cuts

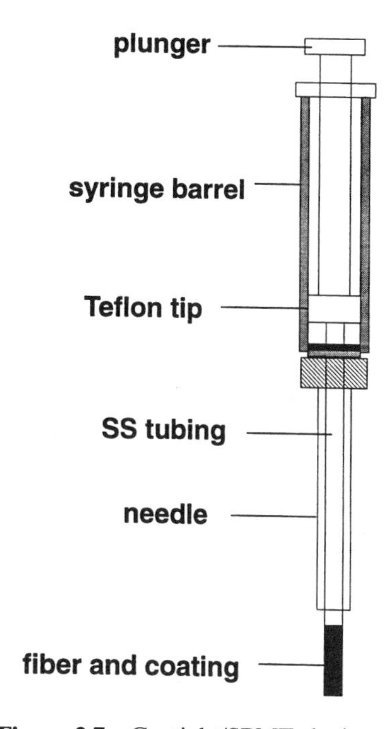

Figure 2.7 Gastight/SPME device.

Source: Adopted with permission from ref. 12.

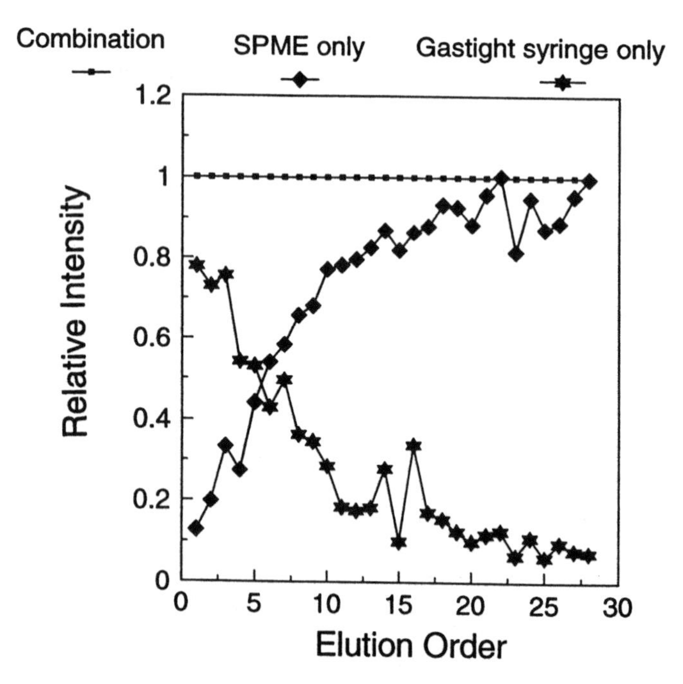

Figure 2.8 Normalized peak area of 28 VOCs, which were sampled from the headspace of a salt-saturated water sample at 100 ppb using three extraction methods; the peak area of VOCs sampled by the gastight/SPME was normalized to 1 for all analytes. The *x* axis gives the elution order numbers of VOCs. (see Table 2.1 for compound identities).

Source: Adopted with permission from ref. 12.

extraction time. The other option is to increase the temperature: this decreases the amount extracted at equilibrium but it may be acceptable if target limits of detection still can be reached.

To enable the simultaneous analysis of very volatile substances (gases) and less volatile analytes, the headspace SPME technique can be combined with static headspace sampling by means of the gastight/SPME device illustrated in Figure 2.7.[12] A fused silica fiber, coated with PDMS, is connected with 30 gauge stainless steel (SS) tubing. The empty end of this SS tubing/fiber assembly is then mounted to the plunger of a Hamilton 500 μL gastight syringe. When the fiber is withdrawn into the syringe needle, a certain volume of gas is also withdrawn into the gastight syringe through the needle opening. For sampling, the fiber is first withdrawn into the syringe needle, which punches through the sample vial septum. The fiber is then exposed into the headspace by lowering the plunger for a predetermined period of time to establish analyte equilibrium among the coating, the headspace, and the sample matrix. Movement of the plunger up and down can be used to increase mass transport from the headspace to the coating and shorten the equilibration times. Then

Table 2.1 The Elution Order of 28 VOCs, Their Boiling Points, and the Estimated Detection Limits for Gastight/SPME Method.*

Elution Order	Compound Name	Boiling Point (°C)	SPME/Headspace LODs (ppt)
1	chloromethane	−24	12.1
2	vinyl chloride	−14	50
3	bromomethane	3.4	74.6
4	chloroethane	12	121
5	trichlorofluoromethane	24	28.3
6	1,1-dichloroethylene	35	7.1
7	methylene chloride	41	23.5
8	trans-1,2-dichloroethylene	48	6.2
9	1,1-dichloroethane	57	32.8
10	trichloromethane	62	36.6
11	1,1,1-trichloroethane	73	12.5
12	tetrachloromethane	77	10.5
13	benzene	80	8.7
14	1,2-dichloroethane	84	58.6
15	trichloroethylene	87	6.3
16	1,2-dichloropropane	97	18.4
17	bromodichloromethane	90	23
18	2-chloroethylvinylether	108	33.9
19	cis-1,3-dichloropropene	104	10.9
20	toluene	112	1
21	trans-1,3-dichloropropene	112	9.7
22	1,1,2-trichloroethane	114	36.8
23	tetrachloroethylene	121	1.7
24	dibromochloromethane	119	16
25	chlorobenzene	132	0.6
26	ethylbenzene	136	0.3
27	bromoform	149	12.6
28	1,1,2,2-tetrachloroethane	142	6

Source: Ref. 12.

the plunger is raised to a premarked position, which allows a predetermined volume of headspace gas to be withdrawn into the gastight/SPME device. The raised plunger also withdraws the fiber into the needle for protection. Upon injection, the volume of headspace is injected at the same time as analytes are desorbed.

Figure 2.8 illustrates the distribution of 28 typical volatile analytes listed in Table 2.1, between a PDMS coating of about 1 μL volume and the 110 μL headspace contained in the sampling device from Figure 2.7. The presence of the gaseous headspace in the gastight SPME device improves sensitivity only for very volatile analytes with boiling temperatures below room temperature. The resulting limit of detections (LODs) are mid to low parts per trillion (ppt) for the full range of volatiles (see Table 2.1).

Figure 2.5c shows the principle of indirect SPME extraction through a membrane. The main purpose of the membrane barrier is to protect the fiber against damage, similar to the use of headspace SPME when very dirty samples are analyzed. However, membrane protection is advantageous for determination of analytes having volatility too low for the headspace approach. In addition, a membrane made from appropriate material can add a certain degree of selectivity to the extraction process. The kinetics of membrane extraction are substantially slower than for direct extraction, though, because the analytes need to diffuse through the membrane before they can reach the coating. Use of thin membranes and increased extraction temperatures will result in faster extraction times.[13]

2.4 Coatings

Equation 2.1 and 2.2 indicate that the efficiency of the extraction process is dependent on the distribution constant K_{fs}. This is a characteristic parameter that describes properties of a coating and its selectivity toward the analyte versus other matrix components. Specific coatings can be developed for a range of applications. Coating volume determines method sensitivity as well (see eq. 2.1), but thicker coatings result in longer extraction times (refer to Chapter 3). Therefore, it is important to use the appropriate coating for a given application. This is clearly demonstrated in Figure 2.9, which compares the performance of two different coatings for analysis of polar and nonpolar compounds from an aqueous matrix. The distribution constant and the sensitivity of the method drop over two orders of magnitude for *o*-xylene, and increase by an order of magnitude for 2,4-dichlorophenol when the film is changed from nonpolar PDMS (Figure 2.9a) to polar poly(acrylate) polymer (PA) (Figure 2.9b).[14] Coating selection and design can be based on chromatographic experience. For example, a very pronounced difference in selectivity toward target analytes and interferences can be achieved by using surfaces common to affinity chromatography.

To date, several experimental coatings have been prepared and investigated for a range of applications. In addition to liquid polymeric coatings such as PDMS for general applications, other more specialized materials have been developed. For example, ion exchange coatings were used to remove metal ions and proteins from aqueous solutions,[15,16] liquid crystalline films to extract planar molecules and carbowax for polar analytes,[2] metal rods to electorodeposit analytes,[11,17] pencil "leads" to extract pesticides,[18] and Nafion coatings to extract polar compounds from nonpolar matrices.[19]

There are several different methods of depositing coatings onto fibers. The dipping technique typically consists of placing a fiber for a short time in a concentrated organic solvent solution of the material to be deposited. After removal of the fiber from the solution, the solvent is evaporated by drying and the deposited material can be crosslinked.[2] An extension of this method is electrodeposition, which can be used to deposit selective coatings on the

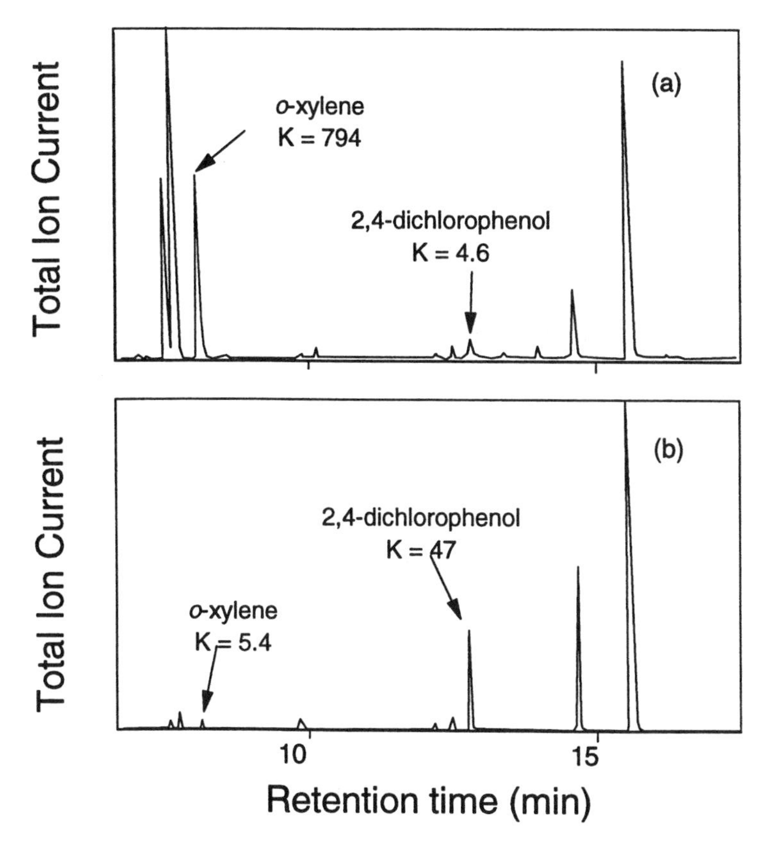

Figure 2.9 Total ion current GC/MS chromatogram of benzene, toluene, ethylbenzene, and *o,m,p*-xylenes (BTEX) and phenol in water extracted with (a) a poly(dimethylsiloxane) coating and (b) a poly(acrylate) coating.

surface of metallic rods.[14] The limitation of this approach is the coating thickness variance from fiber to fiber. Therefore, the preparation of films for commercial devices is carried out simultaneously during the drawing of the fused silica rod. While this process requires dedicated and very expensive equipment, reproducibility of the coating thickness is excellent. This process is identical to the preparation of optical fibers,[20] and so the required equipment is commercially available. Similar results can be obtained by using a piece of hollow fiber membrane (microtubing), made from the desired material. Preparation consists of swelling the membrane by means of an appropriate volatile solvent, placing the enlarged membrane onto the tip of the fiber, and evaporating the solvent. The thickness of the coating is determined by the membrane thickness. A porous hollow fiber membrane can also be used for adsorption of target analytes, or its pores can be filled with organic solvent to allow for solvent microextraction.[21]

2.5 Experimental Parameters that Affect Extraction Efficiency

Salt concentration and sample pH control can be used to enhance extraction; the principle is similar to that for solvent extraction procedures. An appropriate salt or buffer is added directly to the sample without the need for a specific instrument. The other parameter that is very important to optimize is extraction temperature. At elevated temperatures analytes can effectively dissociated native analytes from the matrix and move them into the headspace for rapid extraction by the fiber coatings. However, the coating/sample distribution coefficient also decreases with an increase of temperature, resulting in a diminution in the equilibrium amount of analyte extracted. To prevent loss of sensitivity, the coating can be cooled simultaneously with sample heating. This idea was implemented in the design shown in Figure 2.10. In this device,

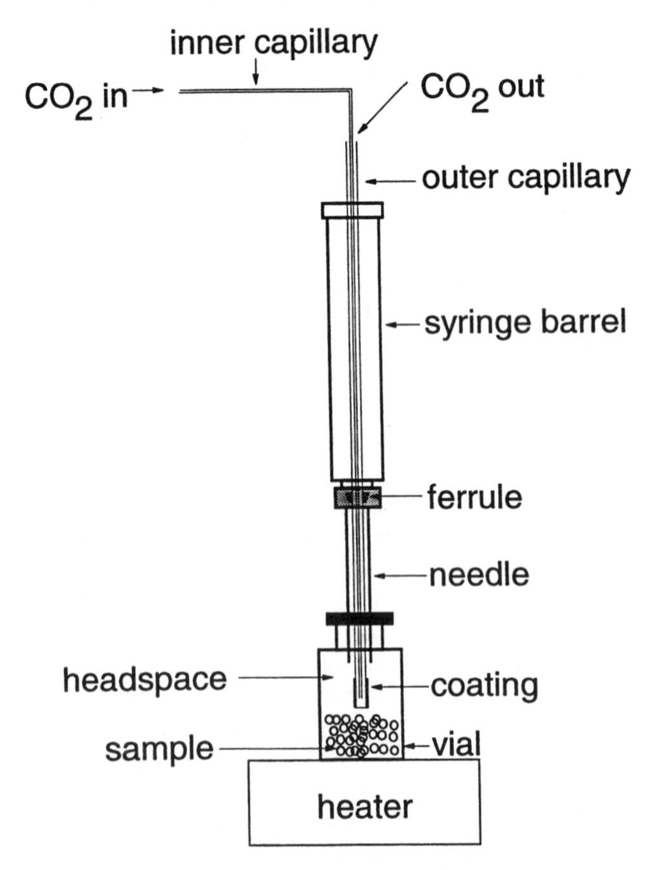

Figure 2.10 Design of internally cooled SPME device.

Source: Adopted with permission from ref. 22.

Table 2.2 Quantitative Extraction of BTEX from Different Matrices

	% Extracted			
Analytes	Water (80°C)	Soil, 15% Water (120°C)	Sand (110°C)	Clay, 5% Water (170°C)
benzene	50	48	84	64
toluene	65	101	105	84
ethylbenzene	85	96	106	91
m,*p*-xylene	90	99	102	95
o-xylene	100	104	100	100

a fused silica tubing is sealed and coated at one end (outer capillary). Liquid carbon dioxide is delivered via the inner capillary to the coated end of the outer capillary resulting in a coating temperature lower than that of the sample. This "cold finger" effect results in accumulation of the analytes at the tip of the fiber. Quantitative extraction of many analytes, including volatiles, is possible with this method.[22]

Table 2.2 illustrates that complete recoveries are possible with the internal cooling approach, even for volatile organic compounds such as substituted benzene, in a range of matrices. This approach can be used effectively to increase the sensitivity of the SPME methods without changing the chemical nature of the fiber. One significant drawback of increased fiber capacity is loss of selectivity, since now not only the analytes but also most of the interferences are extracted exhaustively into the coating.

Water has proven to be a very effective additive to facilitate the release of analytes from the matrix and it is often used to accelerate extraction.[22] It can also be used in combination with high temperature extractions to remove and dissolve even very nonpolar analytes such as PAHs. This is possible because the dielectric constant of water decreases rapidly with temperature increase.[23] This property can also be used also in the solid phase microextraction of solids. Figure 2.11 shows the dynamic hot water extraction system. The solid sample is placed into the extraction vessel which is in the form of a piece of thick-walled tubing. This is important to facilitate efficient removal of the extracted components from the extraction cell.[24] The vessel is then fitted into a high pressure system consisting of a pump delivering water, a heater to control extraction temperature, and a restrictor to maintain the extraction pressure. Water with extracted analytes is collected in a cooled vial supplied with agitation and an SPME device (see Figure 2.11). The analytes are extracted into the fibre coating from the water immediately upon delivery to the vial. Good quantitation has been obtained for extraction of PAHs from contaminated soil when the collection vial is cooled and the SPME fiber is immersed in water continuously during the extraction.[25]

Figure 2.12 shows how the extraction system can be made simpler and less expensive: when a static extraction technique is performed in a high pressure

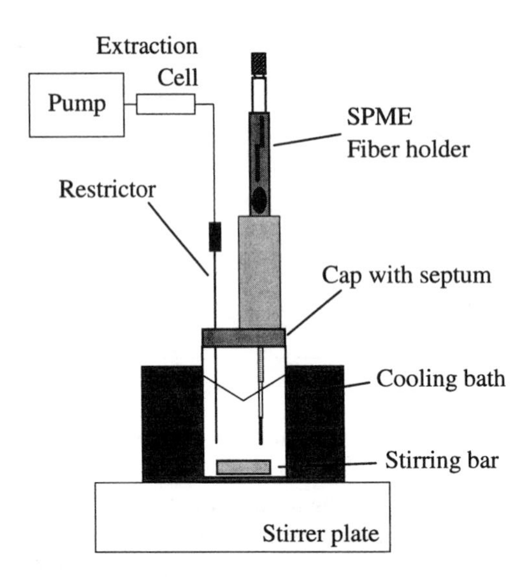

Figure 2.11 Hot water dynamic extraction system combined with SPME for analysis of solid samples.

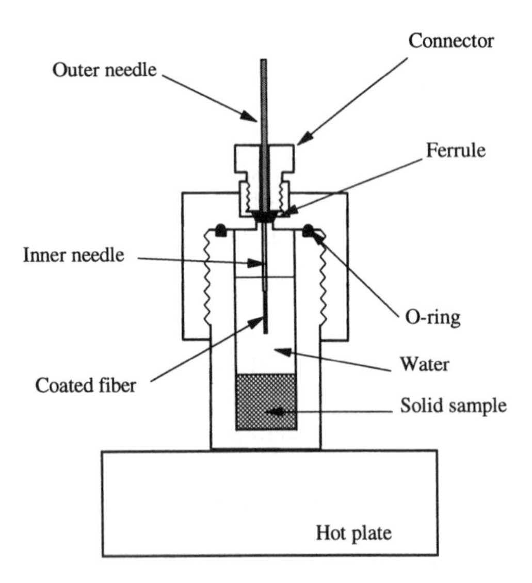

Figure 2.12 Static hot water/SPME system.

vessel, the need for a high pressure pump is eliminated. The procedure consists of adding the sample and water to the vial, inserting the fiber, and sealing it. The extraction can be performed in either direct, headspace, or membrane protection mode (see Figure 2.5). The vial is heated to facilitate the release of analytes to water. After a cooling period to ensure good partitioning onto the fiber, the extracted analytes are desorbed into the analytical instrument and typically quantified, based on an isotopically labeled standard spike (introduced to the sample prior to extraction).

2.6 Derivatization

The main challenge in organic analysis is polar compounds. These are difficult to extract from environmental and biological matrices and difficult to separate on the chromatographic column. Derivatization approaches are frequently used to address this challenge. Figure 2.13 summarizes various derivatization techniques that can be implemented in combination with SPME.[26] Some of the techniques, such as direct derivatization in the sample matrix, are analogous to well-established approaches used in solvent extraction. In the direct technique, the derivatizing agent is first added to the vial containing the sample. The derivatives are then extracted by SPME and introduced into the analytical

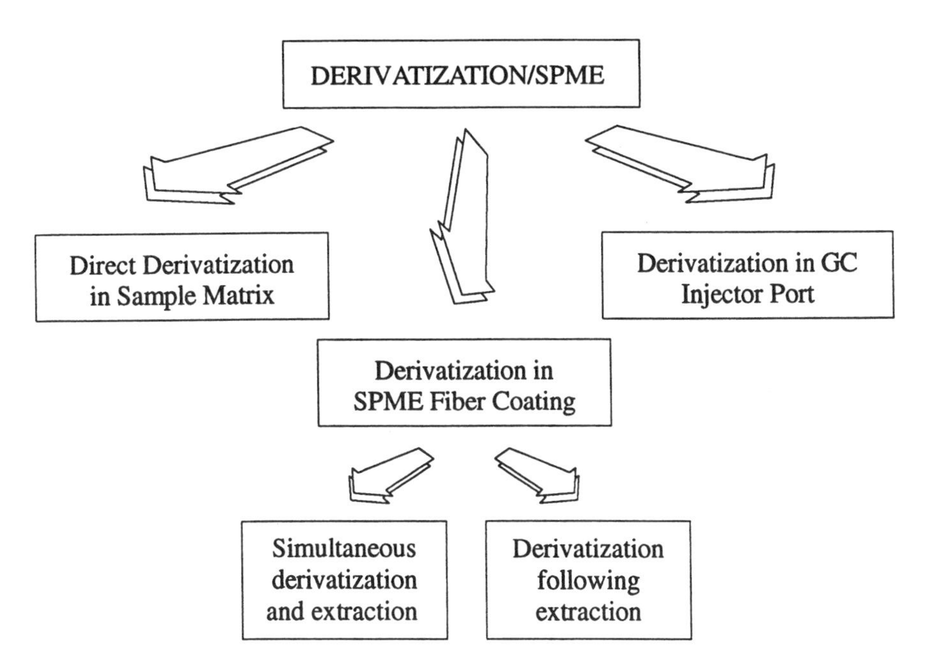

Figure 2.13 SPME/derivatization techniques.

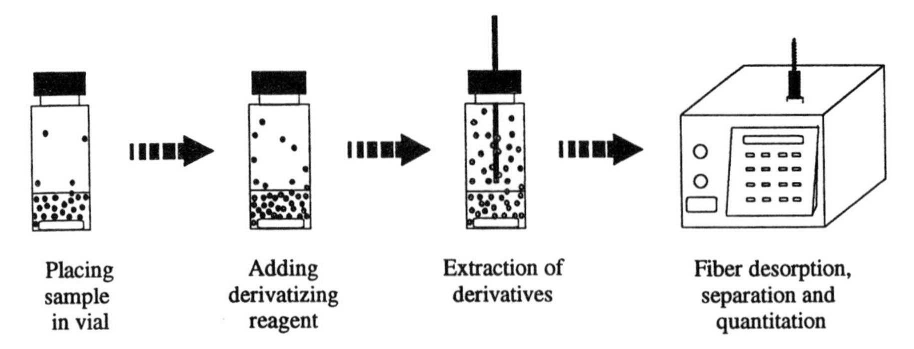

Figure 2.14 In-sample matrix direct derivatization/SPME technique.

instrument (Figure 2.14). For example, this approach has been applied to extract and separate phenols from aqueous samples by first converting the target analytes to their acetate derivatives.[27]

Because of the availability of polar coatings, extraction efficiency for polar underivatized compounds is frequently sufficient to reach the sensitivity required. Frequently, however, there are problems associated with the separation of these analytes. Good chromatographic performance and detection can be facilitated by in-coating derivatization following extraction. This has been shown with high molecular weight carboxylic acids. After exposing to diazomethane, an SPME coating containing extracted analytes, the resulting ester derivatives can be separated as narrow bands on a GC column (see Figure 2.15). In addition, selective derivatization to analogues containing high detector response groups will result in enhancement in sensitivity and selectivity of detection. Derivatization in the GC injector is an analogous approach, but it is performed at high injection port temperatures. For example, long chain carboxylic acids can be extracted onto the coating as ion pairs when tetramethylammonium hydrogen sulfite is added to the sample. During volatilization, analytes are converted to methyl esters.[26]

The most interesting and potentially very useful technique is simultaneous derivatization and extraction, performed directly in the coating. This approach allows high efficiencies and can be used in remote field applications. The simplest way to execute the process is to dope the fiber with a derivatization reagent and expose it to the sample. Then the analytes are extracted, converted to analogues having high affinity for the coating. This is no longer an equilibrium process as derivatized analytes are collected in the coating as long as extraction continues (see Figure 2.16). This approach, which is used for low molecular weight carboxylic acids, results in exhaustive extraction of gaseous samples.[28] When 1-pyrenyldiazomethane is used as the derivatization reagent, it is introduced into the coating by first dissolving the reagent in a volatile solvent. The fiber is then immersed in the solution. The fiber coating swells and is doped with the reagent. After evaporation of the solvent, the fiber is

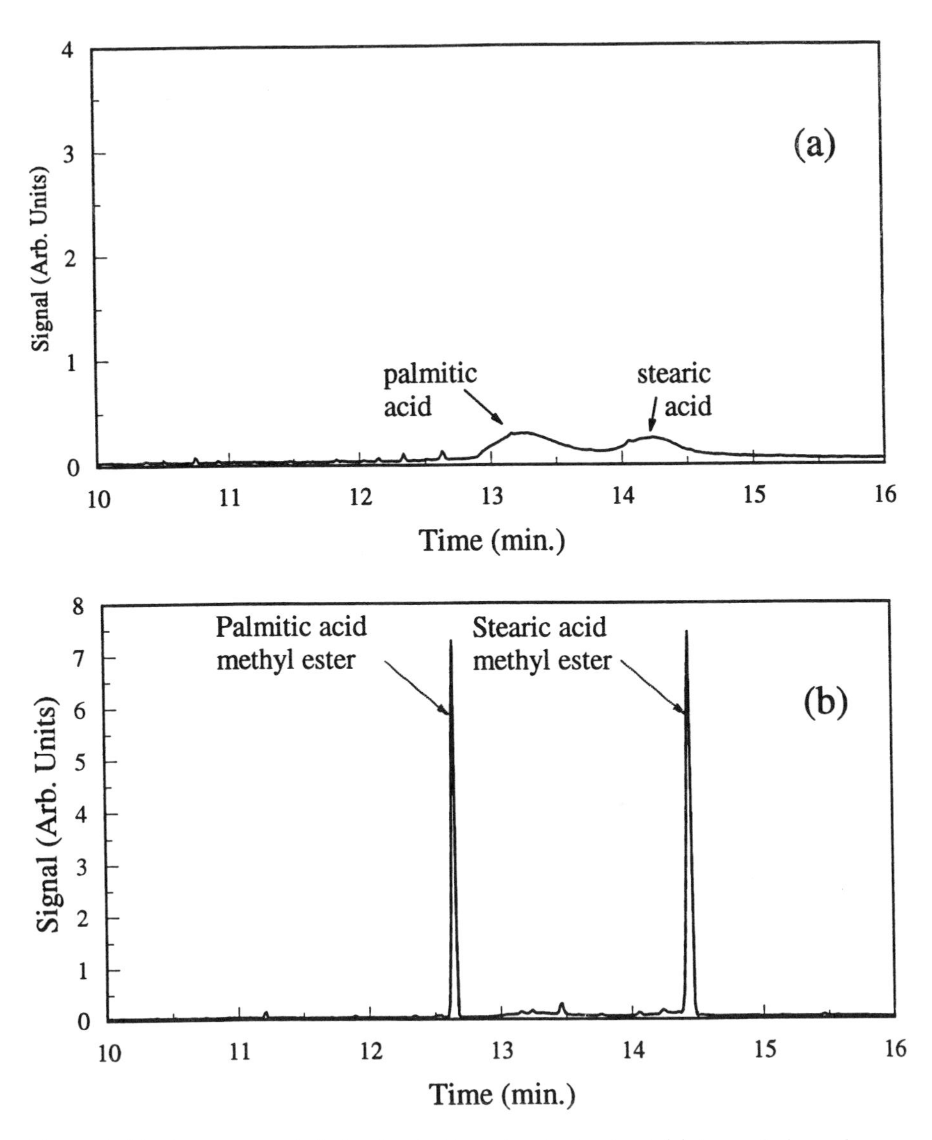

Figure 2.15 Separation of derivatized (b) and underivatized (a) long chain carboxylic acids.

ready to perform extraction. The reagent, having low vapor pressure and high affinity toward the coating, remains on the fiber during its exposure to the sample. Volatility of the pyrenylmethyl esters formed during the reaction also are low, resulting in the accumulation of the product onto the fiber until analyte or reagent is exhausted or decomposed. At high injector temperature the derivatized analytes are removed from the coating and the fiber can be

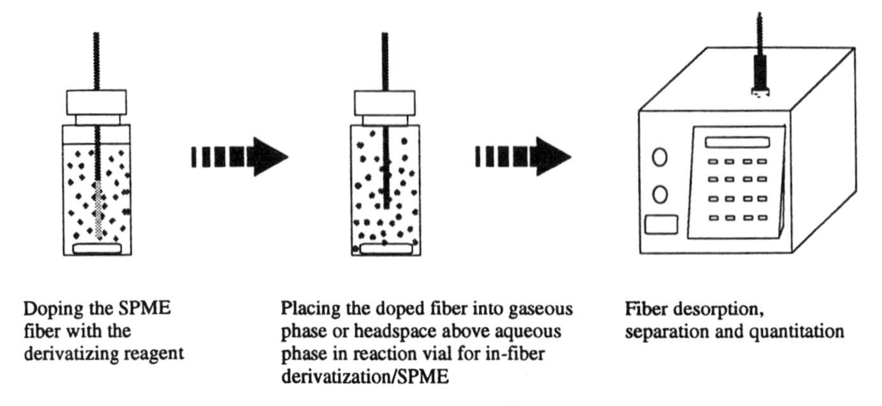

Figure 2.16 In-coating derivatization technique with fiber doping method.

reused. This simple, but powerful procedure is limited to low volatility reagents. The approach can be made more general by chemically attaching the reagent directly to the coating. The chemically bound product can than be released from the coating either by high temperature in the injector, light illumination, change of the applied potential etc. The feasibilty of this approach

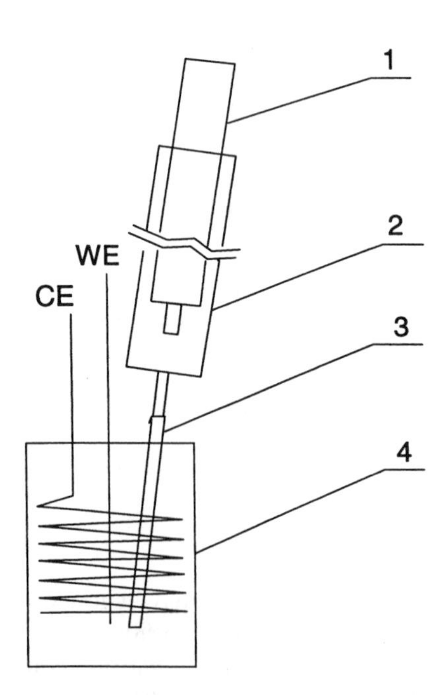

Figure 2.17 SPME/electrochemistry minicell: 1, reference electrode; 2, 100 mL plastic syringe; 3, Teflon capillary; 4, 8 mL Teflon vial; 5, SPME fiber working electrode (WE); 6, platinum wire counter electrode (CE).

was recently demonstrated by synthetizing bouned standards to the silica gel, which were released during heating. This approach allowed solvent-free calibration of the instrument[29]. The other interesting option is to use an SPME device similar to the one shown in Figure 2.10, but the derivatization reagent, instead of carbon dioxide, is delivered to the coating.

In addition to using a chemical reagent, electrons can be supplied to produce redox processes in the coating and convert analytes to more favorable derivatives. In this application, the rod as well as the polymeric film must have good electrical conductivity. Figure 2.17 shows the schematic of the three-electrode cell used to deposit mercury species onto a gold-coated metallic fiber.[17,14] A similar principle has been used to extract amines onto a pencil "lead" electrode.[30]

2.7 Interfaces to Analytical Instrumentation

Because of its solvent-free nature, SPME can be interfaced conveniently to analytical instruments of various types. Only extracted analytes are introduced into the instrument, since the extracting phase is nonvolatile. Thus there is no need for complex injectors designed to deal with large amounts of solvent vapor, and these components can be simplified for use with SPME. The sensitivity of determinations using the SPME technique is very high, facilitating trace analysis. Although in most cases the entire amount of analytes is not extracted from the sample, all material that is extracted is transferred to the analytical instrument, resulting in good performance. Also, the solvent-free process results in narrow bands reaching the instrument, giving taller, narrower peaks and better quantitation.

The analytical instrument used most frequently with SPME has been the gas chromatograph. Standard GC injectors, such as split/splitless can be applied to SPME as long as a narrow insert with an inside diameter close to the outside diameter of the needle is used. The narrow inserts are required to increase the linear flow around the fiber, resulting in efficient removal of desorbed analytes. The split should be turned off during SPME injection. Under these conditions, the desorption of analytes from the fiber is very rapid, not only because the coatings are thin but because the high injector temperatures produce a dramatic decrease in the coating/gas distribution constant and an increase in the diffusion coefficients. The speed of desorption in many cases is limited by the introduction time of the fiber into the heated zone.

One way to facilitate sharper injection zones and faster separation times is to use rapid injection autosampling devices. An alternative solution is to use a dedicated injector, which should be cold during needle introduction, but which heats up very rapidly after exposure of the fiber to the carrier gas

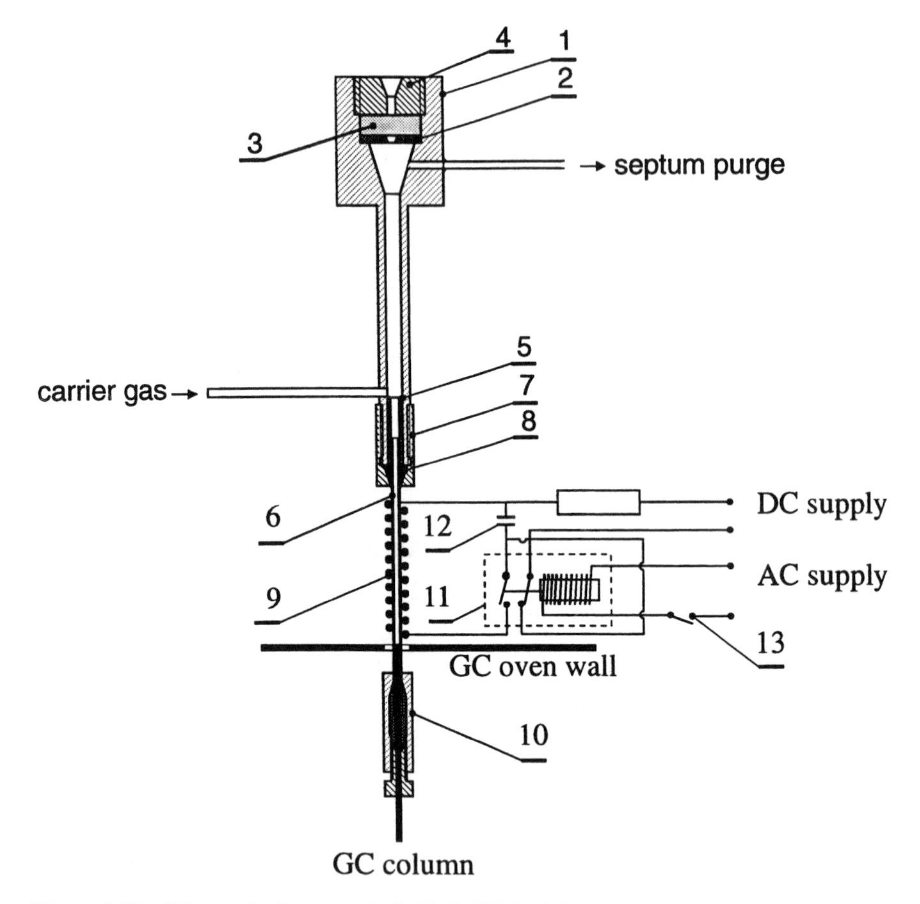

Figure 2.18 Schematic diagram of the flash SPME injector: 1, injector body; 2, washer; 3, septum; 4, nut; 5, needle guide; 6, 0.53 mm i.d. fused silica capillary; 7, nut; 8, ferrule; 9, heater; 10, butt connector; 11, relay; 12, capacitor; 13, switch.

Source: Adopted with permission from ref. 32.

stream. A schematic diagram of such an injector is presented in Figure 2.18. During desorption, the fiber is located inside the heated part of the fused silica capillary, its end being close to the bottom of the heated zone. The distance between the fiber and the capillary wall is approximately 0.15 mm. A close match between inner diameter of the capillary and the outer diameter of the fiber assures effective heat transfer from the heater to the fiber, and a high linear flow rate of the carrier gas along the fiber. The injector is rapidly heated via a capacitive discharge. Heating rates of 1000°C/s have been determined experimentally.[31]

The injector just described has achieved separation of BTEX in 9 seconds (see Figure 2.19). Separation of 28 volatile organic compounds listed in the U.S. EPA method 624 has been accomplished in 150 seconds with reproducibil-

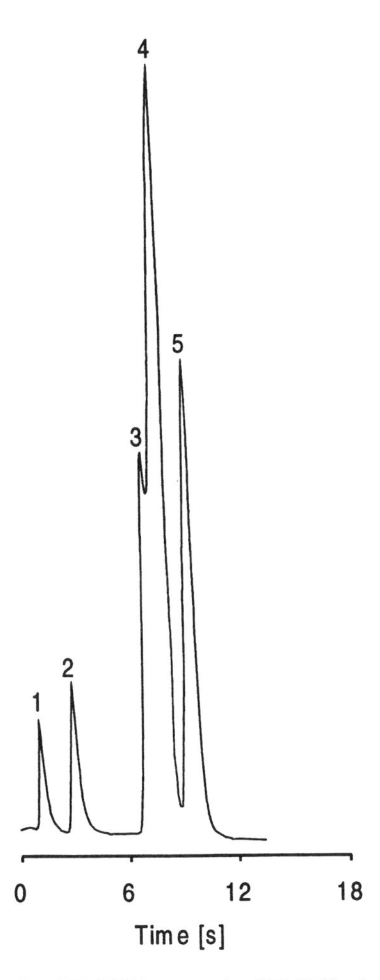

Figure 2.19 Rapid analysis of BTEX in water by SPME/flash injector/GC: 1, benzene; 2, toluene; 3, ethylbenzene; 4, *m*,*p*-xylenes; 5, *o*-xylenes.

ity better than 5% RSD for most analytes.[32] The fiber can be also be designed to contain the heating element, as shown in Figure 2.20. In this case, no injector is necessary. The modified fiber can be introduced directly into the front of the column, and analytes can be desorbed rapidly by heating with a capacitive discharge current after the fiber has been withdrawn from the needle.

Flash desorption injectors can be designed alternatively by passing a current directly through a fiber. This is possible if the rod is made of conductive material, as it is in the case of the electrochemical SPME devices already mentioned. Figure 2.21 illustrates such an interface.[14] When the electrical connection is made at the bottom of the interface, the fiber is rapidly heated

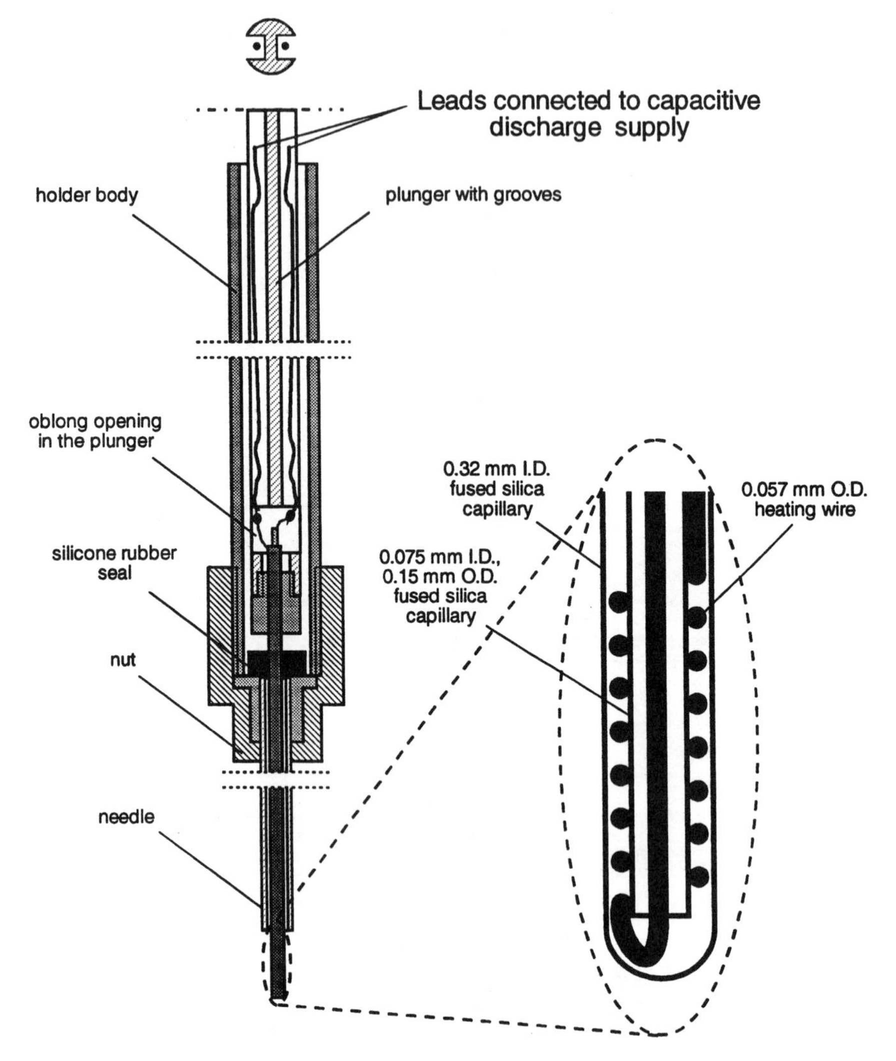

Figure 2.20 Internally heated SPME device.

Source: Adopted with permission from ref. 32.

by the discharging current. The other option is to use laser energy to desorb analytes from the surface of fused silica optical fiber, as discussed at the beginning of this chapter.

Flash desorption injectors can be applied to directly interface SPME to a range of detection devices such as mass spectrometers and atomic emission devices. The sharp bands obtained during the desorption process result in very sensitive detection. For example, Figure 2.22 shows the sharp peak corresponding to about 1 pg of toluene desorbed from the SPME fiber and directly

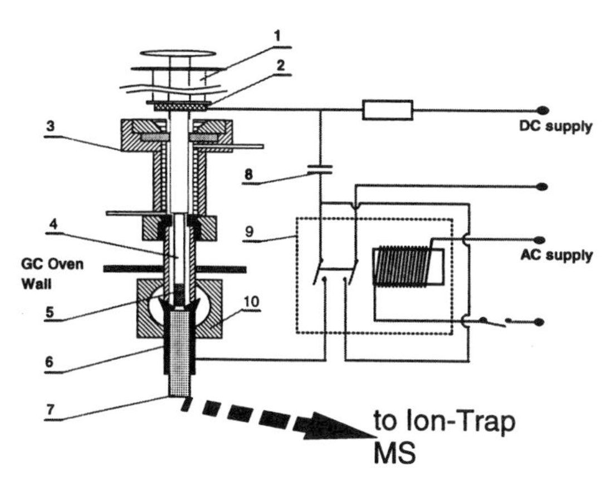

Figure 2.21 Direct capacitive discharge desorption system: 1, SPME syringe; 2, electric connection I; 3, injector body; 4, steel wire; 5, gold coating; 6, electric connection II; 7, transfer line; 8, capacitor; 9, relay; 10, butt connector.

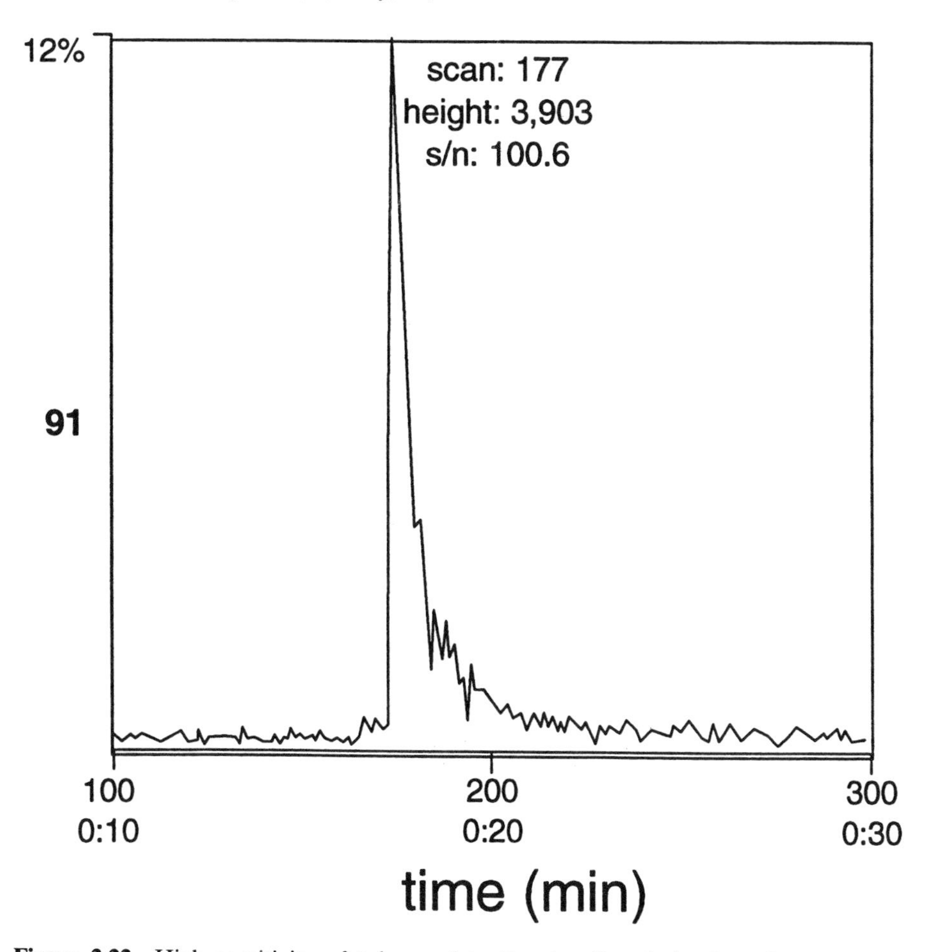

Figure 2.22 High sensitivity of toluene detection by directly hyphenating SPME device to an ion trap mass spectrometer.

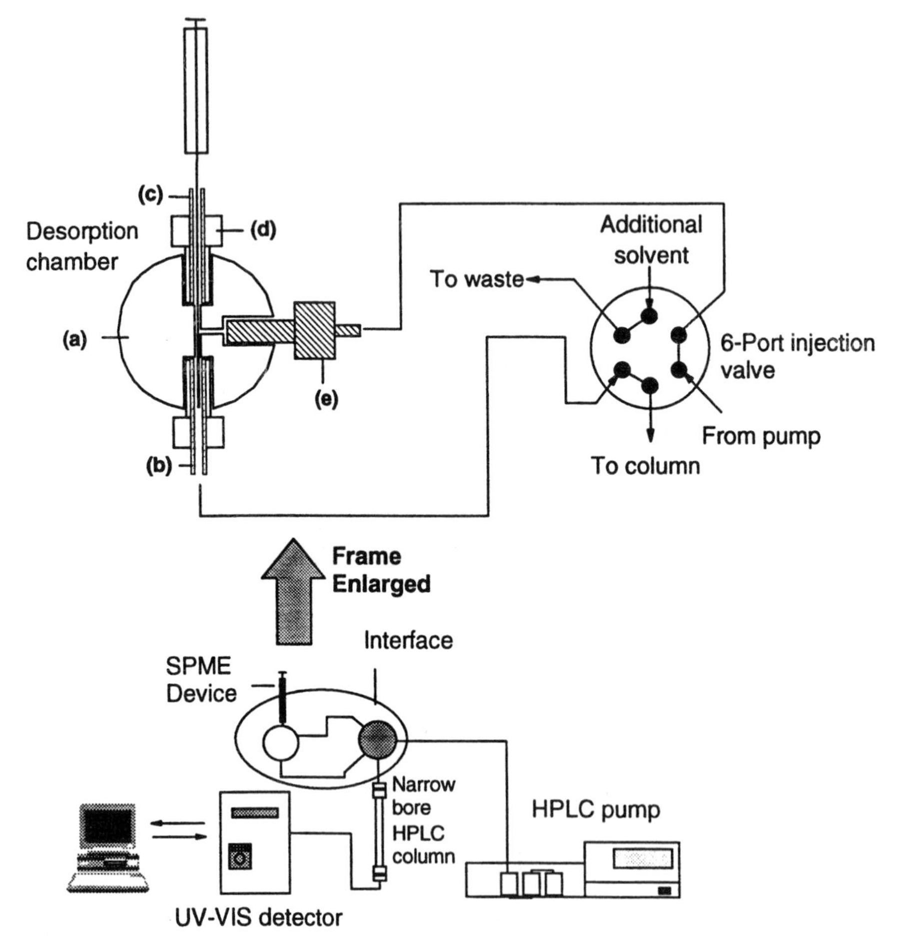

Figure 2.23 SPME/HPLC interface: (a) stainless steel (SS) 1/16 in. tee, (b) 1/16 in. SS tubing, (c) 1/16 in. PEEK tubing (0.02 in. i.d.), (d) two-piece, finger-tight PEEK union, (e) PEEK tubing (0.005 in. i.d.) with a one-piece PEEK union.

detected by the mass spectrometer. The limit of detection is about two orders of magnitude lower than is obtainable with GC/MS techniques, because of a much sharper band. To facilitate proper quantitation, the extract needs to be very clean, which puts an additional demand on the coating selectivity. Some help in proper quantitation can be obtained if the apparatus for tandem mass spectrometry (MS/MS) is available.

Research effort has also been focused on designing interfaces for liquid phase separation techniques to address the need for analysis of nonvolatile and thermally labile analytes. The interface to high performance liquid chromatography (HPLC) can be a straightforward analogue of the traditional loop injection system. A typical SPME/HPLC interface consists of a custom-made

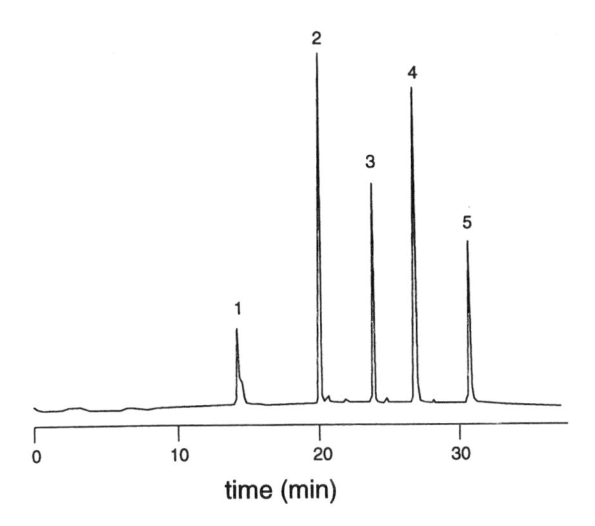

Figure 2.24 Separation of PAHs on a 50 micrometer ID capillary using the SPME device as a sample introduction technique: naphthalene (1), acenaphthene (2), phenanthrene (3), fluoranthene (4) and benz[a]anthracene (5).

Source: Adopted with permission from ref. 35.

desorption chamber and a six-port injection valve (Figure 2.23).[33] The upper part of the PEEK tubing, (c), fitted into a tee-union, is enlarged to fit the needle of the syringe. The internal tubing of the SPME device, which holds the fiber, can be sealed by the PEEK tubing and the tee-union tightly enough to withstand solvent pressures as high as 4500 psi. The desorption chamber is placed in the position at which the injection loop normally resides on the injection valve. When the injection valve is at "load" position, it allows the fiber to be introduced into the desorption chamber under ambient pressure. It also allows for the introduction of a desorption solvent if different from the mobile phase. A heater can be installed in the device to facilitate the desorption process. The interface performs well, and its desorption volume is similar to the volume of the typical injection loop. The smaller injection volumes for applications of SPME to micro-HPLC and capillary electrophoresis can be accomplished by modifying microinjector designs; the sliding injector developed for capillary isotachophoresis is one example.[34] Use of small-volume desorption chambers results in very efficient supercritical fluid chromatography separation (see Figure 2.24) with very narrow bore capillary columns.[35]

SPME can be directly combined to optical detection, based on reflectometric interference spectrometry[36,37,38]. A light beam, passing through an optically transparent fiber coated with transparent sorbing material interacts with absorbed substances through internal reflection. Therefore, if any of the extracted analytes strongly absorb the transmitted light, there is a loss in intensity that can be detected with a simple optical sensor. These devices demonstrate poor sensitivity primarily because it is difficult to find light wavelengths that are specifically adsorbed by the analytes and not by the coating or interferences. In an alternative design, the light can be passed directly through the absorbing polymer which is then cooled to facilitate high sensitivity of determination.[39]

2.8 Commercial Devices

The first commercial version of the laboratory SPME device was introduced by Supelco in 1993 (see Figure 2.25). The device is similar in operation principle to the custom-made device shown in Figures 2.1 and 2.2. Some additional improvements include the adjustment for depth of the fiber with respect to the end of the needle, which allows control of the exposure depth in the injector and extraction vessel. The device incorporates such useful features as color marking of the fiber assemblies to distinguish among various coating types. In addition to standard PDMS coatings of various thickness and PA, Supelco developed new mixed phases based on solid/liquid sorption, such as

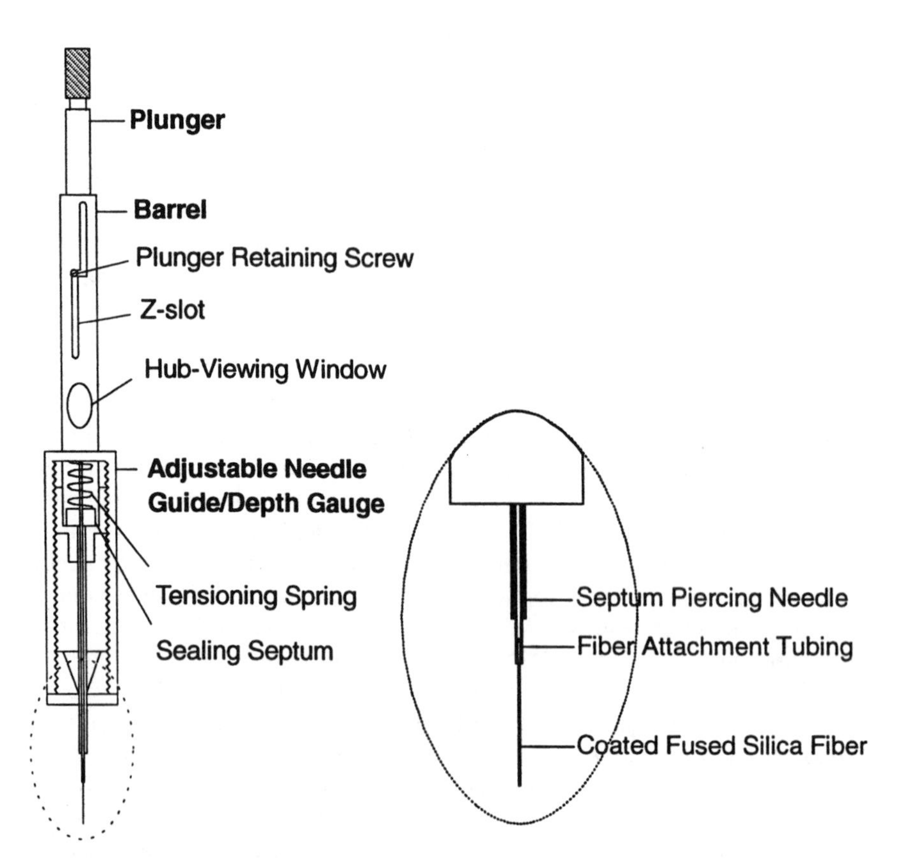

Figure 2.25 Design of the first commercial SPME device made by Supelco.

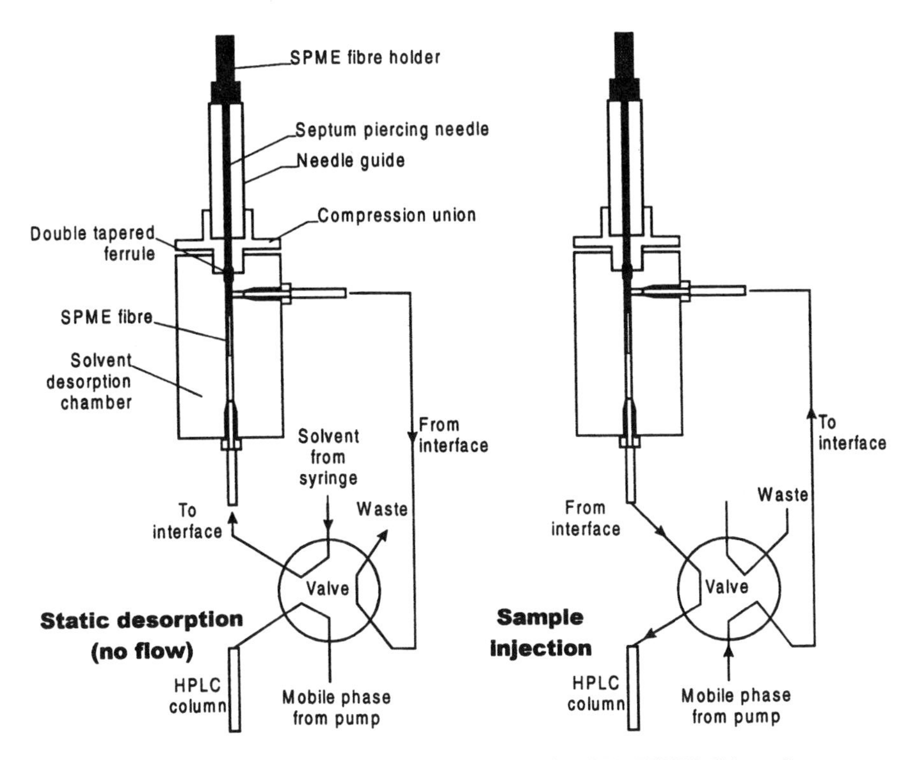

Figure 2.26 Schematic diagram of the Supelco SPME/HPLC interface.

Carbowax/divinylbenzene (DVB) and PDMS/DVB. Supelco also introduced an HPLC interface (Figure 2.26) that integrates the original concept with the injection valve (Figure 2.23).

Varian has been developing the SPME autosampler based on their 8000 GC autosampler system, taking advantage of the fact of the SPME device is analogous to the syringe in its operation and that after desorbtion the coating is cleaned and ready for reuse.[40] The major challenge is to incorporate agitation and temperature control as well as other enhancements, such as an internally fiber cooling or dedicated injectors. One improvement is an SPME system that incorporates an agitation mechanism consisting of a small motor and a cam to vibrate the needle. The vibration causes the vial to shake and the fiber to move with respect to the solution; the result is a substantial decrease of equilibration times compared to a static system. This mode of agitation simplifies fiber handling because it does not require the introduction of foreign objects to the sample prior to extraction. New coatings and devices are expected to follow, as interest in SPME grows along with unprecedented applications.

Small custom modifications of the commercial devices can lead to new possibilities. For example, the addition of input and output connections to

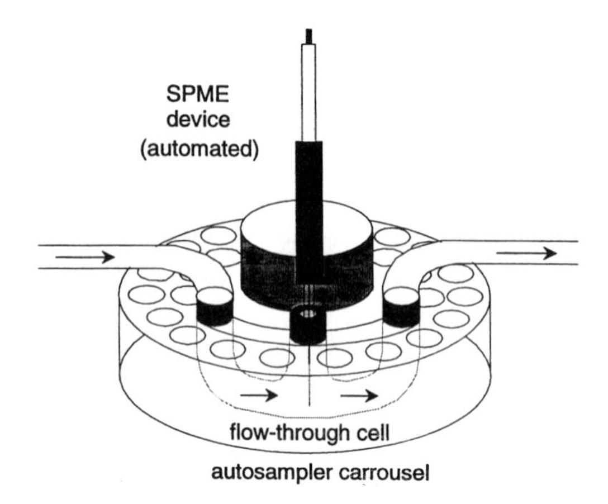

Figure 2.27 On line monitoring of flowing streams by Varian autosampler using the modified vial design.

Source: Adopted with permission from ref. 41.

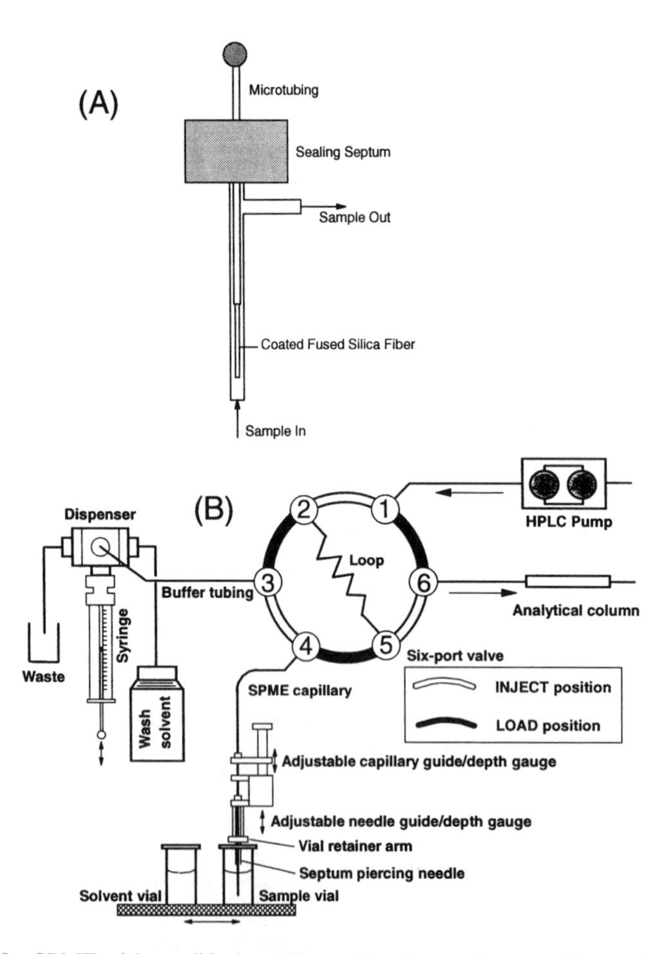

Figure 2.28 SPME with modified needle to allow in-needle extraction and automated flow-through analysis of small samples (A) and automation of an in-tube solid phase microextractor using Spark Holland Autosampler (B).

the autosampler vial allows the system to be used to continuously monitor flowing streams, as shown in Figure 2.27. The flow-through design facilitates agitation of the sample.[41] Alternatively, when a connection is added directly to the needle of the autosampler syringe, as in Figure 2.28A, the system is ready to analyze samples present in the vial.[4] This modified approach relies on air pressure to push the sample through the needle to the vial containing the sample. Then the fiber coating containing the extracted analytes is introduced to the instrument for desorption. Alternatively, the extraction can be performed using a piece of internally coated tubing (a piece of coated capillary column). Figure 2.28B illustrates the system based on a modified Spark Holland micro LC autosampler.[42] In this system the analytes are first extracted into the coating, followed by desorbtion of the compounds using a small volume of solvent. These approaches, which are suitable for analysis of very small samples, offer convenient interfacing to micro HPLC instrumentation.

References

1. J. Pawliszyn and S. Liu, *Anal. Chem.* ***59***, 1475 (1987).
2. R.G. Belardi and J. Pawliszyn, *Water Pollut. Res. J. Can.* ***24***, 179 (1989).
3. C.L. Arthur and J. Pawliszyn, *Anal. Chem.* ***62***, 2145 (1990).
4. J. Pawliszyn, *Method and Device for Solid Phase Microextraction and Desorption,* PCT, International Patent Publication Number WO 91/15745 and national counterparts.
5. M. McComb, E. Giller and H. D. Gesser, in *78th Canadian Society for Chemistry Conference and Exhibition,* (University of Guelph, Guelph, ON, 1995) Abs. 528.
6. M. Chai and J. Pawliszyn, *Environ. Sci. Technol.* ***29***, 693 (1995).
7. C. Grote and J. Pawliszyn, *Anal. Chem.,* ***69***, 587 (1997).
8. D. Louch, S. Motlagh and J. Pawliszyn, *Anal. Chem.* ***64***, 1187 (1992).
9. J. Al-Hawari in *78th Canadian Society for Chemistry Conference and Exhibition,* (Memorial University of Newfoundland, St. Johns, NF, 1996) Abs. 358.
10. S. Motlagh and J. Pawliszyn, *Anal. Chim. Acta* ***284***, 265 (1993).
11. Z. Zhang and J. Pawliszyn, *Anal. Chem.* ***65***, 1843 (1993).
12. Z. Zhang and J. Pawliszyn, *J. High Resolut. Chromatography* ***19***, 155 (1996).
13. Z. Zhang, J. Poerschmann and J. Pawliszyn, *Anal. Com.* ***33***, 219 (1996).
14. Z. Zhang, M. J. Yang and J. Pawliszyn, *Anal. Chem.* ***66***, 844A (1994).
15. E. Otu and J. Pawliszyn, *J. Mikrochim. Acta* ***112***, 41 (1993).
16. J. L. Liao, C. M. Zeng, S. Hjerten and J. Pawliszyn, *J. Microcol. Sep.* ***8***, 1 (1996).
17. F. Guo, T. Gorecki, D. Irish and J. Pawliszyn, *Anal. Commun.,* ***33***, 361 (1996).
18. H.B. Wan, H. Chi, M.K. Wong, C.Y. Mok, *Anal. Chim. Acta* ***298***, 219 (1994).
19. T. Gorecki, P. Martos and J. Pawliszyn, *Anal. Chem.,* submitted.
20. P. Cheo, *Fiber Optics,* Prentice-Hall: Englewood, NJ, 1985, pp. 88.
21. J. Chongrong and J. Pawliszyn, *J. Microcolumn. Sep.,* submitted.

22. Z. Zhang and J. Pawliszyn, *Anal. Chem.* ***67***, 34 (1995).
23. S. Hawthorne, Y. Yang and D. Miller, *Anal. Chem.* ***66***, 2912 (1994).
24. J. Pawliszyn, *J. Chromatogr. Sci.* ***31***, 31 (1993).
25. H. Diamon and J. Pawliszyn, *Anal. Commun.,* ***33***, 421 (1996).
26. L. Pan and J. Pawliszyn, *Anal. Chem.,* ***69***, 196 (1997).
27. K. Buchholz and J. Pawliszyn, *Anal. Chem.* ***66***, 160 (1994).
28. L. Pan and J. Pawliszyn, *Anal. Chem.* ***67***, 4396 (1995).
29. P. Konieczka, L. Wolska, E. Luboch, J. Namiesnik, A. Przyjazny, J. Biernat, *J. Chromatogr.* ***742***, 175 (1996).
30. E.D. Conte and D.W. Miller, *J. High Resolut. Chromatogr.* ***19***, 294 (1996).
31. T. Gorecki and J. Pawliszyn, *J. High Resolut. Chromatogr.* ***18***, 161 (1995).
32. T. Gorecki and J. Pawliszyn, *Anal. Chem.* ***67***, 3265 (1995).
33. J. Chen and J. Pawliszyn, *Anal. Chem.* ***67***, 2530 (1995).
34. T. McDonnell and J. Pawliszyn, *Anal. Chem.* ***63***, 1884 (1991).
35. Y. Hirata and J. Pawliszyn, *J. Microcolumn Sep.* ***6***, 443 (1994).
36. B.L. Wittkamp and D.C. Tilotta, *Anal. Chem.* ***67***, 600 (1995).
37. G.L. Klunder and R.E.Russo, *Anal. Chem.* ***49***, 379 (1995).
38. H.M. Yan, G. Kraus and G. Gauglitz, *Anal. Chim. Acta* ***312***, 1 (1995).
39. J. Pawliszyn, *Device and Process for Increasing Analyte Concentration in a Sorbent,* U.S. Patent 5,496,741.
40. C. Arthur, L. Killam, K. Buchholz, J. Berg and J. Pawliszyn, *Anal. Chem.* ***64***, 1960 (1992).
41. R. Eisert and K. Levsen, *J. Chromatogr.* ***737***, 59 (1996).
42. R. Eisert and J. Pawliszyn, *Anal. Chem., submitted.*

CHAPTER

3

Theory of Solid Phase Microextraction

3.1 Introduction

An understanding of SPME theory both provides insight and direction when developing methods and identifies parameters for rigorous control and optimization. Effective use of the theory minimizes the number of experiments that need to be performed. Theory has been developed to understand the principal processes of SPME by applying basic fundamentals of thermodynamics and mass transfer. To simplify mathematical relationships the theory assumes idealized conditions. Theory for ideal extraction conditions can be very accurate for trace concentrations in simple matrices like air or drinking water at ambient conditions, when secondary factors such as thermal expansion of polymers, changes in diffusion coefficients due to the presence of solutes in polymers, heterogeneity of the matrix and the sorbing phase and so on, can be neglected. When conditions are more complex, theory for ideal cases still approximates well some of the parameters and general relationships between parameters and extraction times or amounts extracted. In this chapter we describe both the thermodynamics and the kinetics of the extraction process. The amount of analyte extracted at equilibrium conditions can be calculated using thermodynamic principles, while the extraction time can be estimated by solving differential equations describing mass transfer conditions in the system.

3.2 Thermodynamics

Solid phase microextraction is a multiphase equilibration process. Frequently the extraction system is complex, as in a sample consisting of an aqueous phase with suspended solid particles having various adsorption interactions with analytes plus a gaseous headspace. Some analysis problems have specific factors to consider, such as biodegradation processes or walls of the sampling vessel adsorbing analytes significantly. To simplify the system, however, only three phases will be considered initially: the fiber coating, the gas phase or headspace, and a homogeneous matrix such as pure water. During extraction, analytes migrate among the three phases until equilibrium is reached. The following discussion is lmited to partitioning equilibrium involving liquid polymeric phases such as PDMS. The method of analysis for solid sorbent coatings is analogues, since the total surface area available for adsorption is proportional to the coating volume if we assume constant porosity of the sorbent.

3.2.1 Multiphase Equilibria in SPME

The mass of an analyte extracted by the polymeric coating is related to the overall equilibrium of the analyte in the three-phase system. Since the total mass of an analyte should remain the same during the extraction as the initial amount, we have:

$$C_0 V_s = C_f^\infty V_f + C_h^\infty V_h + C_s^\infty V_s \tag{3.1}$$

where C_0 is the initial concentration of the analyte in the matrix; C_f^∞, C_h^∞, and C_s^∞ are the equilibrium concentrations of the analyte in the coating, the headspace, and the matrix, respectively; V_f, V_h, and V_s are the volumes of the coating, the headspace, and the matrix, respectively. If we define the coating/gas distribution constant as $K_{fh} = C_f^\infty / C_h^\infty$ and the gas/sample matrix distribution constant as $K_{hs} = C_h^\infty / C_s^\infty$, the mass of the analyte absorbed by the coating, $n = C_f^\infty V_f$, can be expressed as:

$$n = \frac{K_{fh} K_{hs} V_f C_0 V_s}{K_{fh} K_{hs} V_f + K_{hs} V_h + V_s} \tag{3.2}$$

The proper expression for the above distribution constants should involve appropriate activities, but the concentrations are a good approximation considering the trace levels of the analytes in the sample and assuming pure matix. The driving force for multiphase equilibrium is the difference among an analyte's chemical potentials in the three phases. Although SPME can be used to analyze compounds in various matrices and the fiber coating can be immobilized liquid polymers or porous solid materials, in the following text, an analyte in a multiphase system including a liquid polymer, headspace and an aqueous matrix is used for the discussions of the equilibration theory. The general conclusions drawn from the discussions should be valid for other types of coatings and matrices.

The chemical potential of an analyte in the headspace can be expressed as[1]:

$$\mu_h = \mu^0(T) + RT\ln\left(\frac{p_h}{p^0}\right) \tag{3.3}$$

where μ_h is the chemical potential of the analyte in the headspace; p_h is the vapor pressure of the analyte in the headspace; and $\mu^0(T)$ is the chemical potential of the analyte at standard pressure p^0 (usually $p^0 = 1$ atm) and temperature T. Meanwhile, the chemical potentials of the analyte in the coating and the aqueous matrix can be expressed as:

$$\mu_f = \mu^0(T) + RT\ln\left(\frac{p_f}{p^0}\right) \tag{3.4}$$

$$\mu_s = \mu^0(T) + RT\ln\left(\frac{p_s}{p^0}\right) \tag{3.5}$$

where μ_f and μ_s are the chemical potentials of the analyte in the coating and the aqueous matrix, respectively; and p_f and p_s are the vapor pressures of the analyte in equilibrium with the analyte in the coating and the aqueous matrix, respectively. When the three-phase system is at equilibrium, the chemical potentials of the analyte in all three phases must be equal:

$$\mu_f = \mu_h = \mu_s \tag{3.6}$$

From eqs. 3.3–3.6, we can write

$$p_f = p_h = p_s \tag{3.7}$$

According to Henry's law,[1] we have

$$p_f = K_F C_f^\infty \tag{3.8}$$

$$p_s = K_H C_s^\infty \tag{3.9}$$

where K_F and K_H are Henry's Law constants of the analyte in the liquid polymer coating and the aqueous solution, respectively. Assuming that the ideal gas law $p_h V_h = n_h RT$ (n_h is the number of moles of the analyte in the headspace) is valid for the analyte vapor in the headspace,

$$p_h = C_h^\infty RT \tag{3.10}$$

From eqs. 3.7–3.10, we can easily connect the distribution constants with Henry's constants:

$$K_{fh} = \frac{C_f^\infty}{C_h^\infty} = \frac{RT}{K_F} \tag{3.11}$$

$$K_{hs} = \frac{C_h^\infty}{C_s^\infty} = \frac{K_H}{RT} \tag{3.12}$$

In the case of the direct SPME sampling from an aqueous solution, we have $\mu_f = \mu_s$ or $p_f = p_s$ at equilibrium. The distribution constant of the analyte, K_{fs}, between the coating and the aqueous solution can be expressed as $K_{fs} =$

$C_f^\infty / C_s^\infty = K_H / K_F$ since $p_f = K_F C_f^\infty$, $p_s = K_H C_s^\infty$, and $p_f = p_s$ when equilibrium is reached. It is intuitive based on eqs. 3.11 and 3.12 that

$$K_{fs} = \frac{K_H}{K_F} = K_{fh} K_{hs} = K_{fg} K_{gs} \tag{3.13}$$

since the fiber/headspace distribution constant, K_{fh} can be approximated by the fiber/gas distribution constant K_{fg} and the headspace/sample distribution constant, K_{hs}, by the gas/sample distribution constant, K_{gs}, if effect of moisture in the gaseous headspace can be neglected. Then, eq 2 can be rewritten as:

$$n = \frac{K_{fs} V_f C_0 V_s}{K_{fs} V_f + K_{hs} V_h + V_s} \tag{3.14}$$

The equation states, as expected from the equilibrium conditions, that the amount of analyte extracted is independent of the location of the fiber in the system. It may be placed in the headspace or directly in the sample as long as the volume of the fiber coating, headspace, and sample are kept constant. There are three terms in the denominator of eq. 3.14 which gives a measure of the analyte capacity of each phase: fiber ($K_{fs} V_f$), headspace ($K_{hs} V_h$), and sample itself (V_s). If we assume that the vial containing sample is fully filled with the aqueous matrix (no headspace), the term $K_{hs} V_h$ in the denominator, which is related to the capacity ($C_h^\infty V_h$) of the headspace, can be eliminated resulting in:

$$n = \frac{K_{fs} V_f C_0 V_s}{K_{fs} V_f + V_s} \tag{3.15}$$

Both eqs. 3.14 and 3.15 describe the mass absorbed by the polymeric coating after equilibrium has been reached. For most analytes, K_{hs} is relatively small (e.g., benzene has a K_{hs} value of 0.26), and sampling from the headspace will not affect the mass absorbed by the coating if the volume of the headspace is much lower than that of the aqueous solution ($V_h \ll V_s$). The detection limits of headspace SPME are therefore expected to be very similar to those of the direct SPME for these conditions.

The relationship $K_{fs} = K_{fh} K_{hs}$ in headspace SPME can be generalized for multiphase equilibration systems (heterogeneous matrices). If there are n phases (e.g., different solids) present other than the coating and matrix during extraction, and, for convenience, they are numbered from 2 to n starting from the one closest to the coating and ending at the one next to the matrix, the distribution constant between the coating and the matrix ($K_{fs} = C_f^\infty / C_s^\infty$, where C_f^∞ and C_s^∞ are the analyte's equilibrium concentrations in the coating and matrix, respectively) can be expressed as:

$$K_{fs} = K_{f1} K_{12} K_{23} \cdots K_{n-1,n} K_{ns} = K_{f1} K_{ns} \prod_{i=1}^{i=n-1} K_i \tag{3.16}$$

where $K_{f1} = C_f^\infty / C_1^\infty$, $K_{i,i+1} = C_i^\infty / C_{i+1}^\infty$, and $K_{ns} = C_n^\infty / C_s^\infty$ are the distribution constants of coating/1st phase, ith-phase/$i+1$st phase, and nth-phase/matrix. The mass of an analyte extracted by the coating from that matrix is

$$n = \frac{K_{fs}V_fC_0V_s}{K_{fs}V_f + K_{1s}V_1 + K_{2s}V_2 + \cdots + K_{ns}V_n + V_s} \tag{3.17}$$

$$= \frac{K_{fs}V_fC_0V_s}{K_{fs}V_f + \sum_{i-1}^{i=n} K_{is}V_i + V_s}$$

where $K_{is} = C_i^\infty / C_s^\infty$ is the distribution constant of the analyte between the ith phase and the matrix of interest, which can be similarly determined through eq. 3.15. Equation 3.17 turns into eq. 3.15 when there is no intermediate phase during extraction, and into eq. 3.14 when there is a headspace present.

Some real samples are heterogeneous and consist of many immiscible phases. The ability of the fiber coating to extract an analyte, as eq. 3.17 suggests is closely related to (1) the distribution constant (K_{fs}) of an analyte between that particular sample matrix and the coating, which is independent on the number of phases existing during extraction (refer to eq. 3.16), and (2) the capacities of the other phases present in the sample for retaining the analyte. If the capacities of those in-between phases are small (such as for headspace), the mass of the analyte extracted by the fiber coating will not be significantly affected. On the other hand, addition of appropriate phase having high affinity towards interferences but not towards target analytes, can be used to remove unwanted compounds from the fiber coating/sample matrix system. However, it should be emphasized that if those in-between phases are liquid and the analyte has low diffusion coefficients in them, the mass transfer may be slow and the extraction process may be kinetically limited. In addition, the rate of release of analyte from solid matrices in the system can also control the overall mass transfer between the phases.[2,3]

SPME may be used to investigate the distribution of species in multiphase systems, both at equilibrium, and prior to equlibrium in order to study the kinetics of the partitioning process. This is possible primarily because the fiber can be selected to extract only an insignificant portion of the target analytes in a given phase and therefore not affect the distribution in the system. Thus, based on external calibration, it will give information about the concentration of these species in a the phase of interest. Sections 5.8.2 and 5.8.3 discuss the application of SPME to investigate equilibria in water/humic acid mixtures.

3.2.2 Prediction of Distribution Constants

In many cases the distribution constants present in eqs. 3.2–3.17 which determine sensitivity of SPME extraction can be estimated from physicochemical data and chromatographic parameters. For example, distribution constants between a fiber coating and gaseous matrix (e.g., air) can be estimated using isothermal GC retention times on a column with stationary phase indentical to the fiber coating material (see Section 5.8.1).[4] This is possible because the partitioning process in gas chromatography is analogous to the partitioning process in solid phase microextraction, and there is a well-defined relationship correlating distribution constant and the retention times. The nature of the gaseous phase does not affect the distribution constant, unless the components of the gas such as moisture, swell the polymer, thus changing its properties. The formula which correlates the distribution constant and the retention time can be described as:

$$K_{fh} = K_{fg} = (t_R - t_A)F\frac{T}{T_m}\frac{p_m - p_w}{p_m}\frac{3}{2}\frac{(p_i/p_0)^2 - 1}{(p_i/p_0)^3 - 1}\frac{1}{V_L} \tag{3.18}$$

where t_R and t_A are the retention times of the solute and a nonsorbed compound, respectively, F is the column flow measured by a soap-bubble flow meter, T and T_m are the temperatures of the column and flow meter, p_m and p_w are the flow meter pressure and the saturated water vapor pressure, p_i and p_0 are the inlet and outlet pressures of the column, and V_L is the column stationary phase volume. Usually p_m and p_0 are equil to atmospheric pressure. The K_{fg} estimated by this method for the PDMS to gas partitioning of benzene agreed to within a few percent to the estimate by SPME experimentation (see Section 5.8.1).

A most useful method for determining coating-to-gas distribution constants uses the linear temperature programmed retention index (LTPRI) system, which indexes compounds' retention times relative to the retention times of *n*-alkanes. This system is applicable to retention times for temperature-programmed gas-liquid chromatography. The logarithm of the coating-to-air distribution constants of *n*-alkanes can be expressed as a linear function of their LTPRI numbers. For PDMS this relation is log K_{fg} = 0.00415*LTPRI − 0.188.[5] Thus the LTPRI system permits interpolating the curve of K_{fg} versus retention time. LTPRI numbers are available from published tables, so this method can estimate K_{fg} values accurately without experimenting. If the LTPRI number for a compound is not available from published sources, it can be calculated from a GC run according to its definition:

$$LTPRI = (100 \times N) + \left[100 \times \frac{t_{R(A)} - t_{R(N)}}{t_{R(N+1)} - t_{R(N)}}\right] \tag{3.19}$$

where N is the number of carbon atoms for $t_{R(N)}$, $t_{R(A)}$ is the analyte retention time, $t_{R(N)}$ is the n-alkane retention time less than $t_{R(A)}$, and $t_{R(N+1)}$ is the *n*-alkane retention time greater than $t_{R(A)}$. Note that the GC column used to determine LTPRI should be coated with the same material as the fiber coating.

Estimation of the coating water distribution constant can be performed using eq. 3.13, as discussed in more detail in Section 5.2. The appropriate coating/gas distribution constant can be found by applying techniques discussed above, and the gas/water distribution constant (Henry constant) can be obtained from physicochemical tables or can be estimated by the structural unit contribution method.[6]

Some correlations can be used to anticipate trends in SPME coating/water distribution constants for analytes. For example, a number of investigators have reported the correlation between octanol/water distribution constant K_{ow} and K_{fw}. This is expected, since K_{ow} is a very general measure of the affinity of compounds to the organic phase. It should be remembered, however, that the trends are valid only for compounds within the same group having similar structures, such as hydrocarbons, aromatics or phenols; it cannot be used to make comparison between different groups of the compounds, because of different analyte activity coefficients in the polymer (see Sections 5.2 and 5.8). The other approach, which is more universal, is to use the common fragment constants applied for a given coating phase based on theories developed for and tested by liquid chromatography.[7,8,9] In this method, contributions of various functional groups present in a given molecule are added together to estimate appropriate constant.

3.2.3 Effect of Extraction Parameters

Thermodynamics theory predicts the effects of modifying certain extraction conditions on partitioning and indicates parameters to control for reproducibility. The theory can be used to optimize the extraction conditions with a minimum number of experiments and to correct for variations in extraction conditions, without the need to repeat calibration tests at the new conditions. For example, SPME analysis of outdoor air may be done at ambient temperatures that can vary significantly. The equation that predicts the effect of temperature on the amount extracted will allow calibration without the need for extensive experimentation. Extraction conditions that affect K_{fs} include temperature, salting, pH, and organic solvent content in water.

Temperature. If both sample and fiber change temperature from T_0 to T the distribution constant changes according to

$$K_{fs} = K_0 \exp \frac{-\Delta H}{R}\left(\frac{1}{T} - \frac{1}{T_0}\right) \tag{3.20}$$

where K_0 is the distribution constant when both fiber and sample are at temperature T_0 (in kelvin), ΔH is the molar change in enthalpy of analyte when it moves from sample to fiber coating, and R is the gas constant.[10] The enthalpy change, ΔH, is considered constant over temperature ranges typical for SPME experiments. It can be determined by measuring K_{fs} at two different temperatures. For coating/gas distribution constants, ΔH for a volatile compound is well approximated by the heat of vaporization of the pure compound ΔH_v for PDMS.[5,11] Temperature effect must be considered when temperature variations occur while sampling outdoors, and when heating is used to increase extraction rates, stop metabolic activity, or enhance the release of analytes.

When the K_{fs} value for an analyte is greater than 1, the analyte has a lower potential energy in the fiber coating than in the sample, so the analyte partitioning into the fiber must be an exothermic process (i.e. giving off heat), which means ΔH is greater than 0. Therefore eq. 3.20 shows that raising the temperature will decrease K_{fs}.

Equation 3.20 applies only to partitioning between two homogeneous phases. The equation does not apply to partitioning between a fiber and a multicomponent sample, but can still be used to calculate the effect. For example, for PDMS extraction of benzene from the headspace above an aqueous solution, at 25°C $K_{fw} = 125$, $K_{fh} = 493$, ΔH_{fw} has been estimated as 13.9 kJ/K · mol,[5] and ΔH_{fw} can be estimated by the enthalpy change in vaporization (42.8 kJ/K · mol for benzene), where the subscript w represents water and the subscript h represents headspace. Therefore a 100 μm PDMS fiber extracting benzene from 3 mL of headspace above a 2 mL, 100 ppb aqueous solution will extract 5.5 ng at 25°C, according to eq. 3.14. The amount extracted at 90°C can be calculated from eqs. 3.20 and 3.14:

$$n = \frac{n_0}{1 + \frac{1}{V_f} V_w K_{fw} \exp\left[\frac{\Delta H_{fw}}{R}\left(\frac{1}{T} - \frac{1}{T_0}\right)\right] + \frac{1}{V_f} V_h K_{fh} \exp\left[\frac{\Delta H_{fh}}{R}\left(\frac{1}{T} - \frac{1}{T_0}\right)\right]} \tag{3.21}$$

which predicts 0.75 ng. Equation 3.20 does not apply directly to the adsorption process.

Salting. Two common techniques to enhance extraction of organics from aqueous solutions are salting and pH adjustment. Salting can increase or decrease the amount extracted, depending on the compound and salt concentration, and the effect of salting on SPME has been determined to date only by experiment. In general, the salting effect increases with increase of compound polarity. Figure 3–1a illustrate the effect of salting on extraction of benzene and toluene from aqueous matrix. Substential increase of analyte extraction occurs at salt concentrations above 1% and leads to about an order of magnitude increase in sensitivity at 30% level. Saturation with salt can be used to not only lower the detection limits of determination, but also to normalize random salt concentration in natural matrices. Note that salting can lower pH at high salt concentration level, since proton activity is increased with increased solution ionic strength.

The effect of salting on SPME has not been examined theoretically, although theory has been developed for the effect of salting on liquid-liquid extraction. Setchenow's equation is

$$K_{fs} = K_0 e^{k_s C_s} = K_0 \left[1 + k_s C_s + \frac{(k_s C_s)^2}{2!} + \frac{(k_s C_s)^3}{3!} + \cdots\right] \tag{3.22}$$

where K_0 is the partition with no salt, k_s is a constant, and C_s is the salt concentration. This equation predicts salt effects at least for low salt concentrations, and sometimes for high concentrations as well. Long and McDevit have summarized the theory and also tabulated most published research on salting-out and salting-in up to 1951 in an extensive review article.[12]

pH. Adjusting the pH of an aqueous solution will change K for dissociable species, assuming that only the undissociated form of the acid or base can be extracted by the fiber coating, according to:

$$K = K_0 \frac{[H^+]}{K_a + [H^+]} \tag{3.23}$$

where K_0 is the distribution constant between sample and fiber of the undissociated form. This relation was confirmed by Yang and Peppard, whose results for extraction of acids are illustrated in Figure 3.1b.[13] As pH decreases, more acid is present in neutral forms which partition into the coating, resulting in higher sensitivity. To obtain the highest sensitivity, pH needs to be two units lower than the pK value corresponding to the acid.

Polarity of Sample Matrix and Coating Material. The presence of an organic solvent in water changes K according to:

$$K_{fs} = 2.303\ K_{fw} \exp\left(\frac{P_1 - P_2}{2}\right) \tag{3.24}$$

where K_{fw} is the distribution constant fror pure water, $P_1 = 10.2$ is the polarity parameter for water, and $P_2 = cP_s + (1 - c)P_1$ is the water/solvent mixture polarity parameter for a solvent of concentration c and polarity parameter P_s.[6] This equation allows prediction of the distribution constants for water heavily contaminated with miscible solvents, assuming that the solvent does not cause the coating to swell. This relationship indicates that the concentration of the solvent must be above 1% to substantially change the properties of water and the distribution constant. Figure 3.1c illustrate the decrease in extacted amount of BTEX into PDMS coating with increase of methanol concentration in anqueous matrix. Variation of organic composition in the matrix can be compensated for by using internal standard calibration techniques.

Compounds have affinity for a phase of similar polarity. Conzen et al. discussed the significance of polarity on fiber extraction of organics from water.[14] The dielectric constant of PDMS varies from 2.6 to 2.8 depending on polymer molecular weight, and the dielectric constant of PA varies from 2.6 to 3.6.[15] These polymer dielectric constants are similar to those of common organic compounds: for example, ε(toluene) = 2.4 and ε(acetic acid) = 4.1.[16] The polymer and organic ε values are very low compared to that of water at room temperature ($\varepsilon(H_2O) = 78$).[9] The trends in dielectric constants indicate high K_{fs} values for typical organics distributed between fiber coating and water.

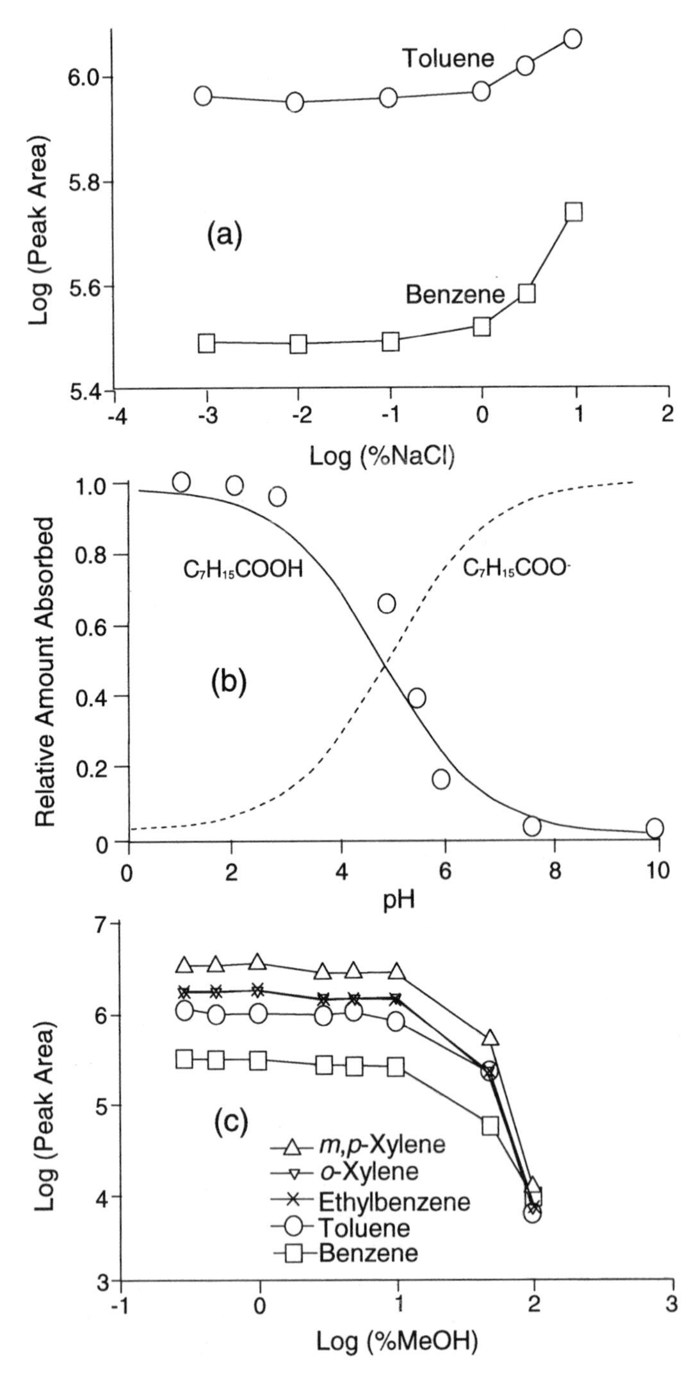

Figure 3.1 The effects of (a) salt (Source: Adopted with permission from ref. 10), (b) pH (Source: Adopted with permission from ref. 13), and (c) solvent (Source: Adopted with permission from ref. 10) on the SPME.

Thermal swelling. The theory developed for SPME assumes ideal conditions and neglects factors such as thermal expansion or interactions between the analytes or with interferences. This idealization is usually justified at ambient conditions and for solutes at trace levels. The effect of thermal expansion on coated fiber extraction was considered by Klunder and Russo.[17] Thermal swelling should cause the fiber coating radius to change according to

$$r = r_0(1 + \alpha T) \tag{3.25}$$

where r_0 is the radius at 0°C, and α is the linear thermal expansion coefficient. For PDMS α is listed as 2.7×10^{-4} °C^{-1}, which is negligible for typical SPME extraction conditions.

3.2.4 Distribution Constants in the Heating-Cooling Environment

High temperature allows the extraction of semivolatile analytes and more efficient release of analytes from the matrix. However, a loss of sensitivity occurs because of the corresponding decrease in the distribution constant. When heating of the sample is combined with simultaneous cooling of the fiber coating, a temperature gap is created between the hot headspace and the cooled fiber coating. This gap provides an additional advantage. In this heating/cooling environment, the coating/headspace distribution constants of analytes increase dramatically.[18]

In headspace SPME, there are two processes involved: the release of analytes from their matrix and the absorption of vaporized analytes by the fiber coating. With the assumption that most analyte molecules can be released into the headspace during extraction, we can simplify headspace SPME into two phases, the fiber coating and the headspace, in the following discussions.

To simplify the theoretical analysis, we consider the extraction of only one analyte, and assume a constant pressure during the extraction process. In reality, the transfer of the analyte from the gas phase to the coating will reduce the pressure, since the volume of the container is kept constant during extraction. But, since we are dealing with trace amounts of analyte, the pressure change during extraction will be negligible.

Headspace SPME using an internally cooled fiber can be described in the following thermodynamic process. At the beginning, a gas sample with an analyte (n_0 moles) and air (n_h moles) is in a container (its volume is V_s) at temperature T_s and pressure P_s, and a liquid coating (n_c moles) is at temperature T_f and pressure P_s. The two systems are isolated from each other. Then, the extraction begins, and the two systems are connected. Only the analyte can be transferred back and forth between the gas phase and the coating. During the extraction, both temperature and pressure of the coating and the gas phase are maintained at their initial values. Finally, the two systems reach a dynamic equilibrium (with n_f moles of the analyte in the coating and n_s moles in the gas phase; $n_0 = n_f + n_s$); that is, the rate of desorption of the

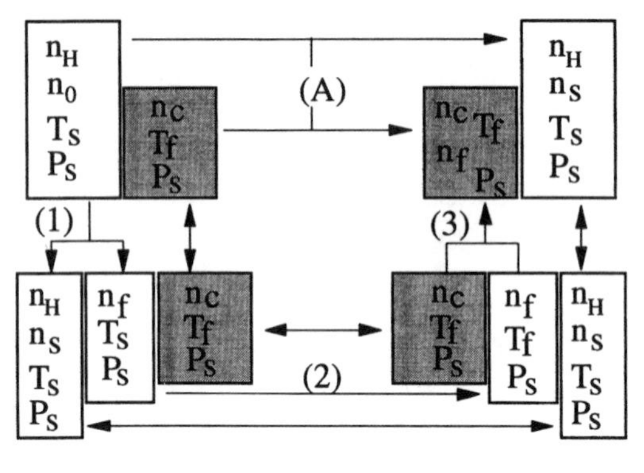

Figure 3.2 Analyte transfer process (A) during internally cooled SPME sampling and its thermodynamic alternative route, steps 1, 2, and 3; n_H, n_c, and n_0 are molar numbers of air, coating materials, and the analyte, respectively; n_s and $n_f(n_s + n_f = n_0)$ are molar numbers of the analyte in the gas phase and the coating, respectively; T_s and T_f are the gas phase and the coating temperatures, respectively; P_s is the pressure of the gas phase; and the two-headed arrow indicates that no change occurs for that phase.

Source: Adopted with permission from ref. 15.

analyte from the coating equals that of the absorption of the analyte by the coating. Route A in Figure 3.2 illustrates this process and the parameters involved.

The driving force of the transfer of the analyte from the gas phase to the coating is the increase in overall thermodynamic entropy consideirng both heat and mass transfer in the system. Equilibrium established when the transfer of an infinitesimal amount of the analyte between the gas phase and the coating does not result in any change in entropy. To calculate the entropy of the extraction process, we can design an alternative route with several simple steps to make the calculation of the entropy easier. The alternative route (steps 1, 2, and 3) is also illustrated in Figure 3.2. The detailed steps of how to calculate the entropy change are discussed in Appendix A. In the following text, only the main equations are discussed.

The entropy change for the first step of the analyte transfer as shown in Figure 3.2 can be expressed as:

$$\Delta S_1 = -n_s R \ln \frac{n_s}{n_H} + n_0 R \ln \frac{n_0}{n_H} \tag{3.26}$$

where n_0 is the number of moles of analyte initially in the headspace; n_s is the number of moles of analyte in the headspace when equilibrium is reached; n_H is the molar number of background gas (mainly air) in the headspace, which is a constant during extraction; R is the gas constant, and it has a value of 8.31 J/K · mol. In the deduction of eq. 3.26, it is also assumed that the gases in the headspace behave ideally and $n_H \gg n_0$.

The second step in the alternative route is the cooling of the part of the analyte gas which is transferred to the coating, from T_s to T_f. The change of the entropy in this step is

$$\Delta S_2 = n_f C_p \left(\frac{T_s - T_f}{T_f} + \ln \frac{T_f}{T_s} \right) = n_f C_p \left(\frac{\Delta T}{T_f} + \ln \frac{T_f}{T_s} \right) \tag{3.27}$$

where n_f is the molar number of the analyte transferred from the headspace to the coating; C_p is the constant pressure heat capacity of the analyte; $T = T_s - T_f$. We assume that the heat capacity of the analyte is constant in the temperature range involved. This assumption can simplify eq. 3.27 considerably without any loss of significant information.

The final step is the absorption of the analyte by the coating at the constant temperature of T_f. The entropy change is

$$\Delta S_3 = -n_f R \ln \left(\frac{T_f n_f V_s}{T_s V_f K_0 n_H} \right) \tag{3.28}$$

The total entropy of the extraction is

$$\Delta S = \Delta S_1 + \Delta S_2 + \Delta S_3 \tag{3.29}$$

As mentioned, when equilibrium is established, the entropy will not change as the result of an infinitesimal transfer of analyte between the gas phase and the coating. In mathematical terms, at equilibrium we have:

$$\frac{\partial \Delta S}{\partial n_f} = 0 \tag{3.30}$$

Keeping in mind the conservation of the analyte $n_0 = n_s + n_f$ while carrying out the partial differentiation of ΔS (eq. 3.29), we can write

$$C_p \left(\frac{\Delta T}{T_f} + \ln \frac{T_f}{T_s} \right) - R \ln \left(\frac{T_f n_f V_s}{T_s n_s V_f K_0} \right) = 0 \tag{3.31}$$

Defining the distribution constant of the analyte between cold fiber and hot headspace as $K_T = C_f/C_s = n_f V_s / n_s V_f$, we then rewrite eq. 3.31 as

$$K_T = K_0 \frac{T_s}{T_f} \exp \left[\frac{C_p}{R} \left(\frac{\Delta T}{T_f} + \ln \frac{T_f}{T_s} \right) \right] \tag{3.32}$$

where $\Delta T = T_s - T_f$; K_0 is the coating/headspace distribution constant of the analyte when both coating and headspace are at temperature T_f.

Equation 3.32 indicates that there are three main elements which affect the distribution constant of a particular analyte in the heating/cooling environment: K_0, the distribution constant of the analyte at temperature T_f, determined by the coating temperature (T_f) and the interaction between the analyte

Table 3.1 K_T Values (Calculated from eq. 3.32 with T_f = 25°C) and Other Parameters of Selected Compounds

T_s (K)	C_p (J/K · mol)	C_p (ave)	K_T/K_0	K_T
		Benzene[a]		
300	83.02	82.73	1.01	496
350		90.36	1.36	670
400	113.52	97.98	2.36	1,161
450		104.44	5.15	2,539
500	139.35	110.9	14.18	6,993
		Toluene[b]		
300	104.42	104.08	1.01	1,330
350		112.96	1.41	1,867
400	139.91	121.83	2.7	3,571
450		129.54	6.92	9,144
500	170.77	137.26	23.57	31,153
		Ethylbenzene[c]		
300	128.19	127.8	1.01	3,286
350		138.24	1.47	4,804
400	169.95	148.68	3.15	10,294
450		157.83	9.65	31,500
500	206.58	166.99	41.77	136,418
		o-Xylene[d]		
300	133.01	132.66	1.01	4,444
350		142.02	1.48	6,537
400	170.46	151.38	3.2	14,139
450		159.85	9.88	43,622
500	204.32	168.32	42.85	189,260

[a] At 298 K, K_0 = 493, C_p = 82.44 J/K · mol.
[b] At 298 K, K_0 = 1322, C_p = 103.75 J/K · mol.
[c] At 298 K, K_0 3266, C_p = 127.4 J/K · mol.
[d] At 298 K, K_0 = 4417, C_p = 132.31 J/K · mol.

and the coating; the coating temperature, T_f; the headspace temperature, T_s. Since the lower the T_f values are, the larger the K_0 is (exothermic absorption), maintaining the fiber coating at a low temperature and creating a large temperature gap can greatly increase the distribution constant. However, there are practical limits that must be considered when applying eq. 3.32. For example, cooling the coating to very low temperatures may affect the coating's physical and chemical properties and reduce the rate of mass transfer within the coating.

From eq. 3.32, distribution constants (K_T) of analytes between the two phases at different temperatures can be calculated. Table 3.1 lists K_T values of BTEX compounds at selected headspace temperatures. Since K_T values for the three xylene isomers are very similar, only *o*-xylene values are listed in the table. Values for C_p values are taken from the Thermodynamic Research Center's tables.[19] During the calculation, the average C_p value (listed as C_p(ave) in Table 3.1) of the compounds in the temperature range is used. The coating is PDMS and T_f = 298K (25°C). The distribution constants of BTEX at T_s = $T_f(K_0)$ are also listed in Table 3.1.

Table 3.1 shows that if the coating temperature is maintained at 25°C as the headspace temperature goes up, the coating/headspace distribution constant of the analyte increases dramatically. For example, toluene has a distribution constant of 1,330 when both the coating and headspace are at 25°C. This value increases to 31,153 when the headspace temperature is at 227°C and the coating temperature at 25°C. With distribution constants known and the assumption that analytes can be completely released from the matrix, the mass of analytes extracted by the coating can be estimated in advance for the headspace extraction by the internally cooled SPME using eq. 3.15. The coating/headspace distribution constant K_{fs} in this equation can be replaced by K_0 if both the coating and headspace are at the same temperature, and by K_T if the coating and headspace are at different temperatures. This dramatic increase frequently leads to exhaustive extraction (see Table 2.2).

3.3 Kinetics

The kinetics of the extraction process determines the speed of extractions. Most of the theory of mass transport is based on Fick's second law of diffusion describing mass balance in a dynamic system, which in one dimension can be expressed as:

$$\frac{\partial C}{\partial t} = D\frac{\partial^2 C}{\partial x^2} \tag{3.33}$$

where C is the concentration and D the diffusion coefficient of the analyte. Considering the cylindrical geometry of the fiber and sampling system in three-dimensional space eq. 3.33 is converted to:

$$\frac{\partial C}{\partial t} = D\frac{1}{r}\left[\frac{\partial}{\partial r}\left(r\frac{\partial C}{\partial r}\right)\right] \tag{3.34}$$

Examining the kinetics based on this equation involves determining values for the diffusion coefficient, determining values for the boundary distribution constants, assuming boundary conditions and solving the differential equations. Methods to solve these problems are well explained in the text *Conduction of Heat in Solids* by Carslaw and Jaeger.[20] The details of derivations and solutions are presented in Appendix A. More informative discussion is presented below using graphs generated from analytical solutions.

The kinetic theory identifies "bottlenecks" of solid phase microextraction and indicates strategies to increase speed of extractions. All diffusion is assumed to behave according to Fick's law. The theory assumes zero interactions between analytes and vial surfaces or fiber core. Factors such as thermal expansion, swelling, and analyte/analyte interactions are assumed to be negligible.

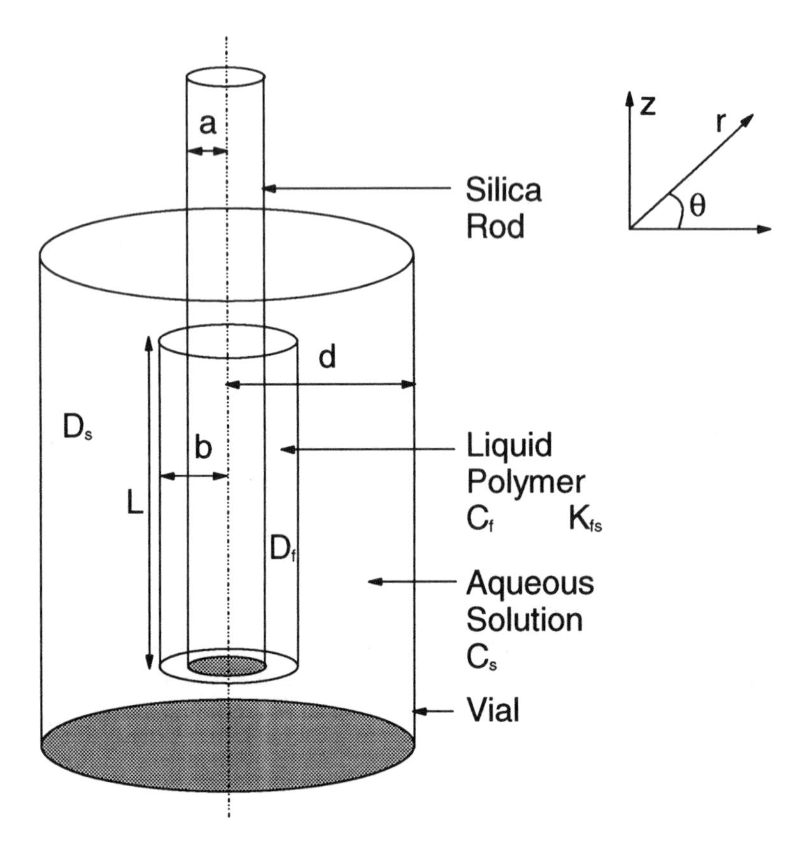

Figure 3.3 Graphic representation of the SPME/sample system configuration, with dimensions and parameters labeled as follows: a, fiber coating inner radius; b, fiber coating outer radius; L, fiber coating length; d, vial inner radius; C_f, analyte concentration in the fiber coating; D_f, analyte diffusion coefficient in the fiber coating; C_s, analyte concentration in the sample; D_s, analyte diffusion coefficient in the sample; K_{fs}, analyte distribution coefficient between fiber coating and sample; $K_{fs} = C_f/C_s$.

Source: Adopted with permission from ref. 25.

3.3.1 Direct Extraction

Let us first examine the direct extraction of the analytes from a homogeneous water sample into a fiber's liquid polymer phase coating. No headspace is present in the system. Figure 3.3 shows the geometry of the system investigated, where $b - a$ is the coating thickness.

Perfect Agitation. Let us first consider the case where the water sample is perfectly aggitated. In other words, the aqueous phase moves very rapidly with respect to the fiber so that all the analytes present in the sample have access to the fiber coating. In this situation the extraction process is shown

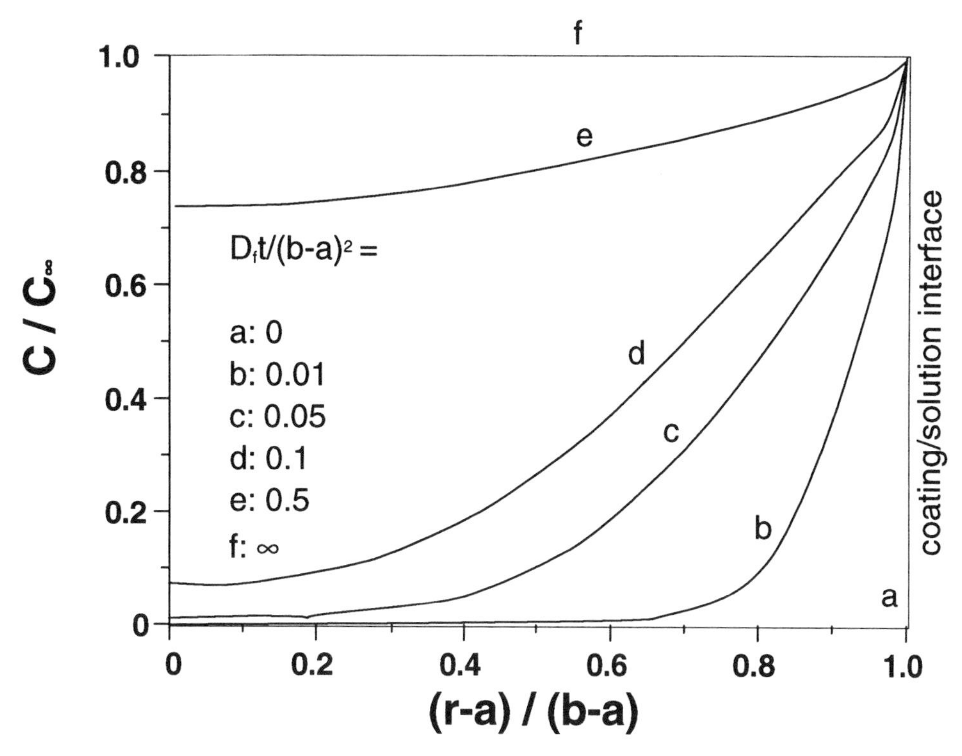

Figure 3.4 Absorption concentration versus radius profiles for different times after the fiber is exposed to a perfectly stirred sample, as calculated by the complete analytical solution given in Appendix A. These profiles are valid for any analyte concentration in the sample, and for any K_{fs} value. The curves have the following values for $Dt/(b - a)^2$: a, 0; b, 0.01; c, 0.05; d, 0.1; e, 0.5; f, infinity.

Source: Adopted with permission from ref. 25.

in Figure 3.4. A series of curves in Figure 3.4 represent concentration profiles obtained at different immersion times of the fiber into the aqueous solution. Both the concentration and position axes use appropriate dimensionless parameters to allow generalization of this graph. The concentration of analyte in the aqueous phase is uniform and constant and is equal to the initial conditions. This constant concentration is ensured by the assumed infinite volume and perfect agitation concentration. In practice, to meet this condition the volume of the sample must be sufficiently large to ensure that the extracted amount does not change within the limits of experimental error with volume increase.

Before the fiber is placed in the solution, no analyte is present in the coating (Figure 3.4a). Immediately after immersion into the sample, only a thin layer close to the surface contains analyte (Figure 3.4b). With time, analyte molecules diffuse progressively deeper into the coating (Figures 3.4c-e) and eventu-

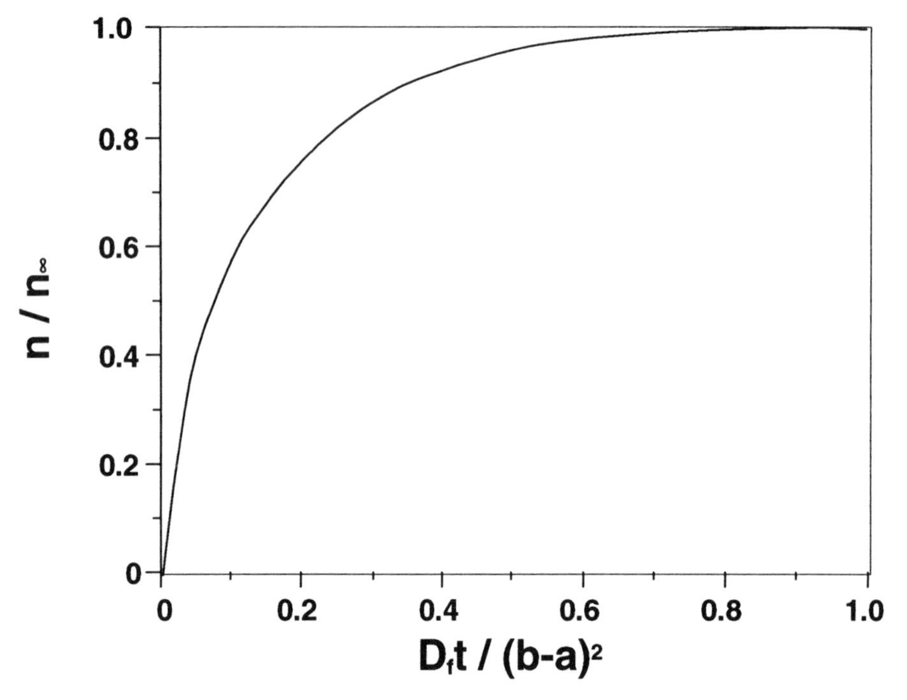

Figure 3.5 Mass absorbed versus time from a perfectly agitated solution of infinite volume. This profile is valid for any analyte concentration in the sample, and for any K_{fs} value.

Source: Adopted with permission from ref. 25.

ally reach equilibrium (Figure 3.4f). Figure 3.4 indicates that at perfect agitation conditions, the speed of the absorption process is determined only by the diffusion of the analyte in the polymer coating.

The area under a concentration profile curve from Figure 3.4 corresponds to the amount of analyte in the coating expressed as a fraction of the mass extracted at equilibrium. This relationship can be shown as in Figure 3.5, and it is referred to as the extraction time profile. This is a universal graph, since both the mass and time axes have dimensionless scales. The graph shows that immediately after immersion of the fiber in the solution there is a rapid increase in the mass absorbed by the fiber. The rate of increase then slows, and eventually reaches equilibrium. The exact formula for extraction from a perfectly agitated,infinite volume, sample is given in Appendix A.

Figure 3.5 illustrates that the time required to reach equilibrium is infinitely long. However in practice a change in mass extracted cannot be determined if it is smaller than the experimental error, which is typically about 5%. Therefore the equilibration time is assumed to be achieved when 95% of the equilibrium amount of an analyte is extracted from the sample:

$$t_e = t_{95\%} = \frac{(b - a)^2}{2D_f} \tag{3.35}$$

Using this equation one can estimate the shortest equilibration time possible for the practical system by substituting appropriate data for the diffusion coefficient of an analyte in the coating (D_f) and the fiber coating thickness ($b - a$). For example, the equilibration time for extraction benzene from a perfectly stirred solution with a 100 μm PDMS film is expected to be about 20 seconds. Equilibration times close to those predicted for perfectly agitated samples have been obtained experimentally for extraction of analytes from air samples (because of high diffusion coefficients in gas) or when very high sonication power was used to facilitate mass transfer in aqueous samples. However in practice there is always a layer of unstirred water around the fiber. A higher stirring rate will result in a thinner water layer around the fiber.

Practical Agitation. Independent of the agitation level, fluid contacting a fiber's surface is always stationary, and as the distance from the fiber surface increases, the fluid movement gradually increases until it corresponds to bulk flow in the sample. To model mass transport, the gradation in fluid motion and convection of molecules in the space surrounding the fiber surface can be simplified by a zone of defined thickness in which no convection occurs

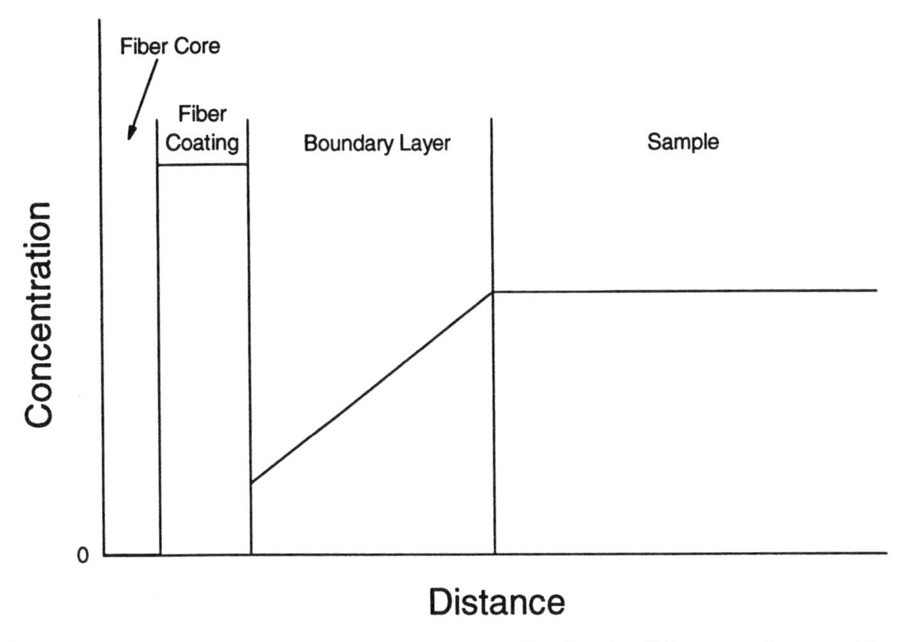

Figure 3.6 Boundary layer model configuration showing the different regions considered and the assumed concentration versus radius profile for when the boundary layer determines the extraction rate.

and perfect agitation in the bulk of the fluid everywhere else. This static layer zone is called a *Prandtl* boundary layer (see Figure 3.6).[21] Its thickness is determined by the agitation conditions and the viscosity of the fluid.

An illustration of this model is presented in Figure 3.7 for the extraction of a hypothetical analyte from water by a 100 μm coating having the following characteristics: $K_{fs} = 5$, $D_s = 1.08 \times 10^{-5}$ cm^2/s and $D_f = 2.8 \times 10^{-6}$ cm^2/s. The series of curves in Figure 3.7 represent concentration profiles obtained at different immersion times of the fiber into the aqueous solution for a thin boundary layer (A, 10 μm) and a thick boundary layer (B, 100 μm) corresponding to well and poorly agitated samples. The concentration of analyte in the aqueous phase outside the boundary layer is uniform and constant and is equal to the initial concentration. This constant concentration is ensured by the assumed infinite volume and perfect agitation conditions in the bulk fluid.

In both cases, before the fiber is placed in the solution ($t = 0$), no analyte is present in the coating (Figure 3.7). Similar to the perfect agitation case (Figure 3.4), immediately after immersion into the sample (5 s), most of extracted analyte is present in a thin layer in the coating close to the surface. However now the concentration in the aqueous phase close to the fiber surface substantially decreases because a concentration profile is also produced in the boundary layer. This results in lower concentration gradients in the coating at the interface and slower mass transport in the system. With time, analyte molecules diffuse progressively deeper into the coating, and eventually equilibrium is reached in the system.

The area under a concentration profile curve in the coating as shown in Figure 3.7 corresponds to the amount of analyte in the coating expressed as a fraction of the mass extracted at equilibrium. This relationship is shown in Figure 3.8 for the thin and thick boundary layers. For comparison, the equilibration time profile for the perfect agitation case is also included. In all cases, immediately after immersion of the fiber in the solution there is a rapid increase in mass absorbed by the fiber which then evens out as system reaches equilibrium. The effect of the boundary layer size on the equilibration rate is very visible. The thin static film does not affect the extraction rate significantly. The equilibration time for a 10 μm thick boundary layer is about 25 seconds (Figure 3.8b), compared to 20 seconds for a perfectly agitated sample (Figure 3.8a). In Figure 3.7c the boundary layer is 100 μm, which is sufficiently thick that diffusion through this zone determines the extraction rate. The equilibration time for a 100 μm thick boundary layer was calculated to be 95 seconds.

The exact formula describing extraction from an agitated sample with a boundary layer is given in Appendix A. Figure 3.9 illustrates the relationship for the case when the extraction rate is determined by the presence of a boundary layer. This is a universal graph, since both the mass and time axes have dimensionless scales. Note that an analyte with a high K_{fs} value will have a long equilibration time even with a very thin boundary layer, characteristic

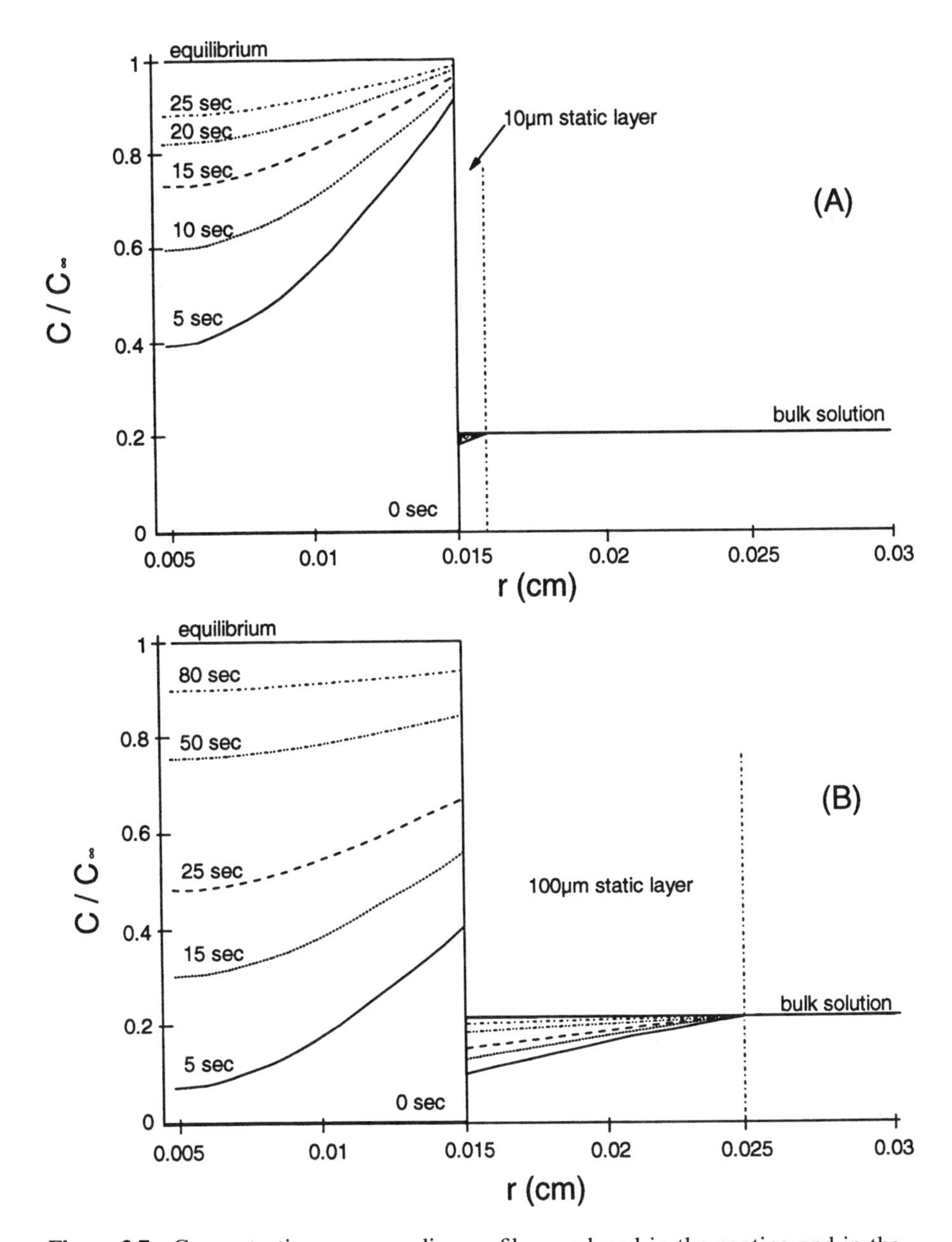

Figure 3.7 Concentration versus radius profiles produced in the coating and in the aqueous phase at different times after fiber is exposed in an SPME fiber/boundary layer system, as calculated by the complete analytical solution given in Appendix A for two cases: (A) boundary layer is 10 mm thick and (B) boundary layer is 100 mm thick. For both cases, the following parameters were used: a = 0.005 cm, b = 0.015 cm, L = 1 cm, $D_s = 1.08 \times 10^{-5}$ cm^2/s, $D_f = 2.8 \times 10^{-6}$ cm^2/s, $K_{fs} = 5$.

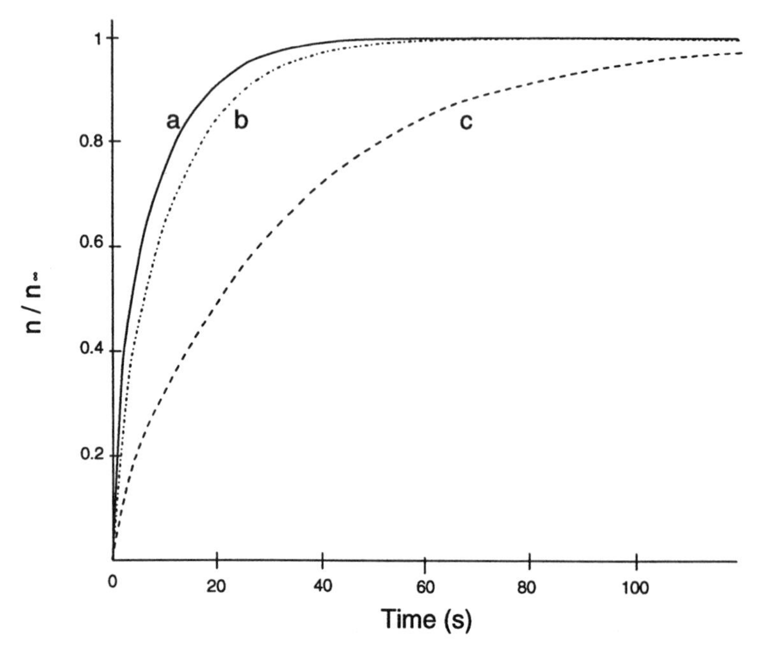

Figure 3.8 Extraction versus time profiles corresponding to the following conditions: a, perfect agitation condition; b, well agitation; c, poor agitation. Experimental parameters same as in Figure 3.7.

of rapid agitation. Similar to the case of eq. 3.35, the time required to reach equilibrium can be estimated from Figure 3.9:

$$t_e = t_{95\%} = 3\frac{\delta K_{fs}(b - a)}{D_s} \tag{3.36}$$

where $(b - a)$ is the fiber coating's thickness, D_s is the analyte's diffusion coefficient in the sample fluid, K_{fs} is the analyte's distribution constant between fiber and sample. This equation can be used to predict equilibration times when the extraction rate is controlled by the diffusion in the boundary layer. In the other words, the extraction time calculated by using eq. 3.36 must be longer than the corresponding time predicted by eq. 3.35, which leads to the following condition:

$$\delta > \frac{(b - a)}{6K_{fs}}\frac{D_s}{D_f} \tag{3.37}$$

Diffusion coefficients for the most common coating, PDMS, are smaller than the corresponding coefficients in water by a factor of 5–6. Therefore the above condition for water extraction and PDMS coating can be simplified to the ratio of the coating thickness and the appropriate distribution constant:

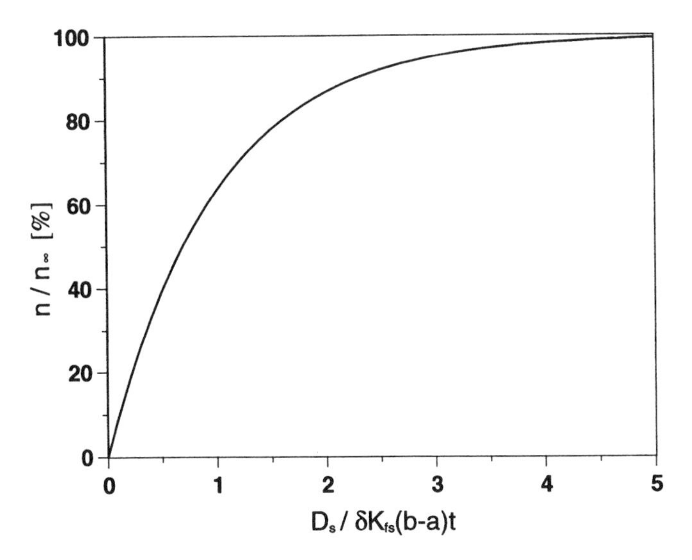

Figure 3.9 Dimensionless extraction versus time profile corresponding to extraction mass absorbed from an agitated solution of infinite volume for when the boundary layer controls the extraction rate.

$$\delta > \frac{b-a}{K_{fs}} \tag{3.38}$$

When the calculated value of $(b - a)/K_{fs}$ is small compared to the thickness of the boundary layer, eq. 3.35 should be used to estimate the extraction time. It should be emphasized that both equations substantially underestimate equilibration times for the situation when their values are close together, since the diffusion through both phases has a cumulative effect on the equilibration times. For better approximation in these cases, the full mathematical expression included in Appendix A should be used.

The effective thickness of a boundary layercan be estimated from empirical formulae of fluid mechanics. If the fiber is placed in the center of a vial stirred by a magnetic bar, liquid flow will be axis-symmetrical around the circumference of the fiber. The equation for a flat plate boundary layer will apply, provided the layer thickness, δ, is small compared to the coated fiber radius, b.[21] From the formula for heat transfer,[22] the effective boundary layer thickness is

$$\delta = 2.64 \frac{b}{\mathrm{Pr}^{0.43}\sqrt{R_d}} \tag{3.39}$$

where b is the radius of the fiber; R_d is the Reynolds number, $R_d = 2ub/\nu$, where u and ν are the fluid's linear speed and kinematic viscosity (for water at 25°C kinematic viscosity is 0.009 cm^2/s), b is the coated fiber radius, and

Pr, the Prandtl number of the liquid, equal ν/D, where D is the diffusion coefficient of the analyte in the liquid.[23] This formula applies for laminar flows: $R_d < 10$.

Extraction time can be shortened by optimizing the position of the fiber in stirred water. If the fiber is placed off-center in the vial so that the fluid flows past the fiber and perpendicular to the fiber axis, the boundary layer thickness can be estimated by:

$$\delta = 9.52 \frac{b}{R_d^{0.62} \mathrm{Pr}^{0.38}} \tag{3.40}$$

The tangential velocity in water agitated by stir bar in a cylindrical container is predicted by

$$u(r) = 1.05\pi N r \left[2 - \left(\frac{r}{0.74R} \right)^2 \right] \quad \text{for } 0 < r \leq 0.74R \tag{3.41}$$

$$u(r) = 0.575\pi N R^2 \frac{1}{r} \quad \text{for } r > 0.74R \tag{3.42}$$

where R is the radius of the stir bar and N the revolutions per second.[24] The point of maximum velocity is $r = 0.56R$, where the velocity is $u(r) = 2.64NR$.

For example, when extracting benzene from water ($K_{fs} = 125$) at 25°C using a 56 μm PDMS coated fiber (b = 125 μm) placed in the center of the vial and magnetic stirring at 1000 rpm, eq. 3.39 predicts the boundary layer thickness of about 10 μm. This thickness satisfies the preceding condition (eq. 3.38: 10 μm > 56 μm/125), which means that the extraction rate is controlled by the diffusion in the boundary layer. Therefore we can use eq. 3.36 to calculate equilibration time, which is found to be 180 seconds. If, on the other hand the fiber is placed at the point of maximum velocity, $r = 0.56R$, for a stir bar 2 cm long ($R = 1$ cm) the same equations predict an equilibration time of 90 seconds. The experimental extraction times are close to these values (see later, Figure 3.31). This calculation indicates that the position of the fiber in the vial should be kept constant, preferably close to the optimum position, about half the distance between the center of the vial and the end of the stir bar. In practice, overestimation of the equilibration times is advisable to eliminate increased extraction as a function of minute change in the fiber position.

The extraction time profile from Figure 3.8 indicates that a 10 μm boundary layer (Figure 3.8b) does not affect the equilibration time significantly compared to the perfect agitation case, while in the discussion above the diffusion through the 10 μm layer controls the kinetics of extraction. The difference is associated with the distribution constant, which in the case of Figure 3.8 is only 5 while in the example above is 125. This comparison illustrates that not only the thickness of the boundary layer is important, but also the amount of analyte which needs to be transported through it to reach equilibrium. Therefore, for extraction from a small volume sample resulting in a drop of analyte concentration in the bulk of the sample, the equilibration time will be shorter since the amount extracted will be smaller.

It should be emphasized that eqs. 3.39 and 3.40 can be used to calculate the boundary layer thicknesses for other agitation methods as long as relative velocity of sample phase versus coating can be estimated. For example, it is possible to estimate the boundary layer thickness for fiber movement agitation methods when the average velocity of the fiber in the solution is known.

Sampling From a Static System. Estimation of the analyte distribution in a system with a perfectly static sample can also be performed using the equations developed for the boundary layer case (Appendix A). In this situation we assume that $\delta = d$, where d is the radius of the vial.

Figure 3.10 presents theoretical curves showing the concentration distribution of compound characterized by $K_{fs} = 100$ in a 100 μm coating and water sample, during extraction from a perfectly static solution. Similar to the perfect agitation case (see Figure 3.4), most of the extracted analyte is present initially in a thin layer of the polymer located close to the water/coating interface. However the concentration of the analyte is lower compared to the well agitated case because the layer of sample close to the fiber is now depleted of analyte and a concentration profile has also formed in the sample in the vicinity of the fiber. Transport of analyte from the progressively thicker depleted layer to the fiber coating determines overall extraction speed. The more analytes need to be transported to the fiber, the longer the process will last. For comparison, Figure 3.11 shows the concentration profiles obtained for the static extraction of the compound characterized by $K_{fs} = 5$. In this case the amount extracted at equilibrium is 20 times lower, resulting in substantially faster equilibration, since the analytes need to be transported only from the vicinity of the fiber.

Figure 3.12 presents equilibration time profiles obtained for the two static extraction cases discussed above. Plot 3.12c corresponds to $K_{fs} = 100$ and a 7 μm thick polymer phase versus the 100 μm used in previous cases. As expected, the appropriate equilibration times are affected by the change of the parameters defining the amount of analytes which need to be transported to the fiber to reach equilibrium. For example, equilibration time for $K_{fs} =$ 100 and 100 μm coating thickness is about 12,000 seconds (see Figure 3.12a), while for $K_{fs} = 5$ it corresponds to only 600 seconds (see Figure 3.12b), which is 20 times shorter, closely matching the ratio of appropriate distribution constants. Similarly, by changing the coating thickness from 100 μm to 7 μm, the amount of analyte which is transported to the fiber during the extraction process dropped over an order of magnitude, resulting in a similar reduction in the equilibration time (compare Figure 3.12a with 3.12c). The shapes of curves from Figures 3.12 b and c are not the same although the equilibration times are similar. The difference reflects the dissimilarity in the coatings' surface-to-volume ratios in each case.

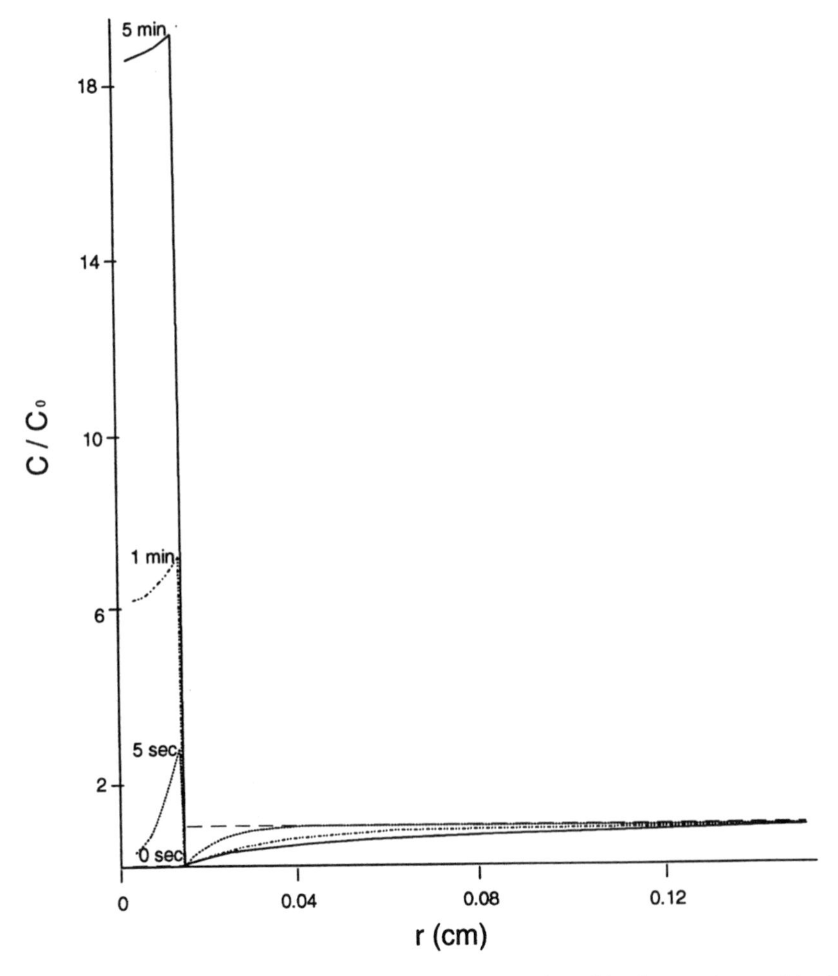

Figure 3.10 Concentration versus radius profiles produced in the coating and in the aqueous phase at different times for extraction from a perfectly static solution, as calculated by the complete analytical solution given in Appendix A. The parameters used were K_{fs} = 100, a = 0.005 cm, b = 0.015 cm, L = 1 cm, D_s = 1.08 × 10^{-5} cm^2/s, D_f = 2.8 × 10^{-6} cm^2/s.

In practice it is very difficult to obtain perfectly static conditions since vial vibration and convection of the fluid associated with the thermal gradient are always present in the system. If very small vials are used (d = 0.5 mm) then the static conditions can be well simulated since the viscosity of the matrix ensures a stationary system. In that case perfect agreement with experiment is obtained.[25]

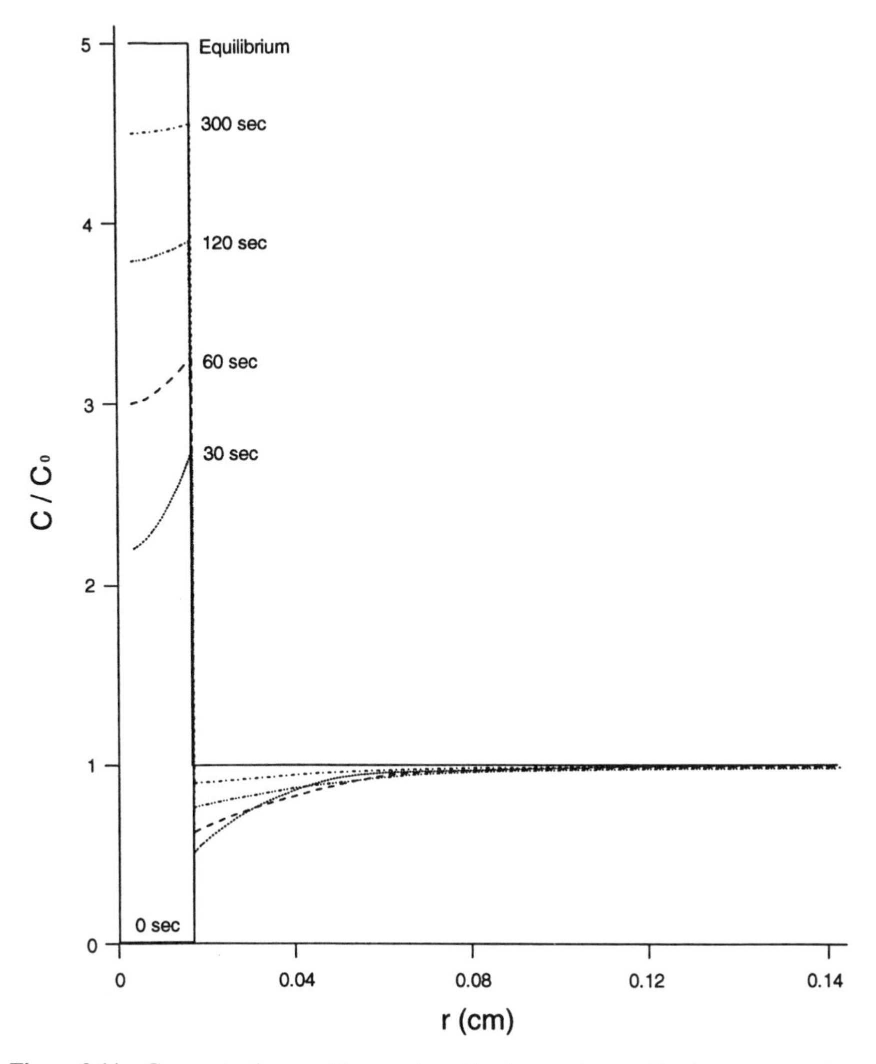

Figure 3.11 Concentration profiles produced in the coating and in the aqueous phase for the conditions and parameters described for Figure 3.10, except $K_{fs} = 5$.

3.3.2 Desorption of Extracted Analytes

After the extraction is complete, the coated fiber containing analytes is transferred to the injection port of a GC or HPLC instrument. During the desorption process, the analyte diffuses from the coating into the stream of carrier

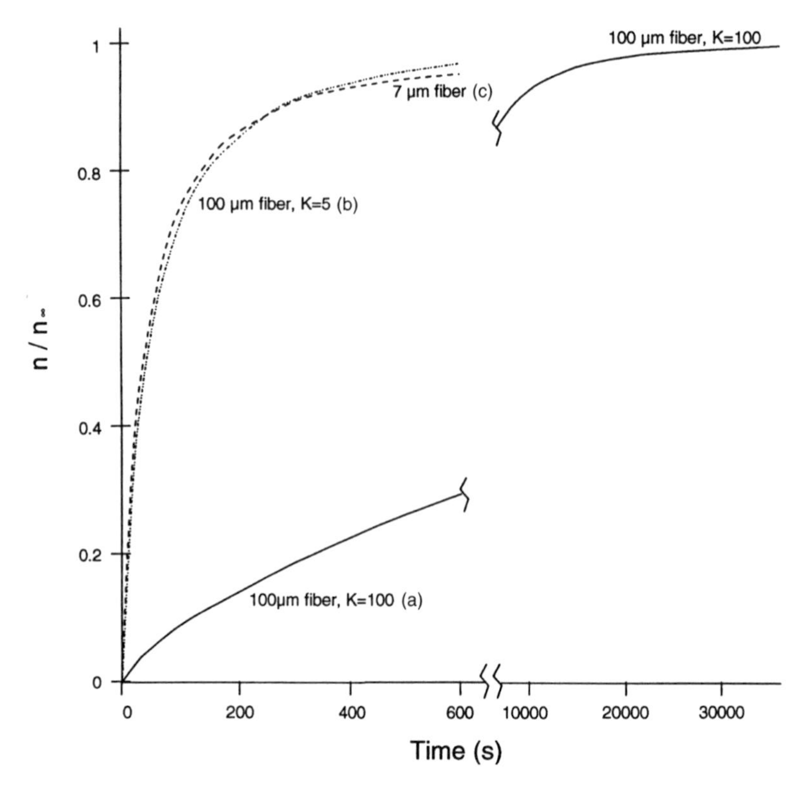

Figure 3.12 Extraction versus time profiles corresponding to (a) Figure 3.10, (b) Figure 3.11 and (c) for a 7 μm thick fiber coating (the same parameters and conditions as described for Figure 3.10 except $a = 0.005$ cm and $b = 0.0057$ cm).

fluid. Therefore, this process is the reverse of absorption from a well-agitated aqueous phase, when the concentration of an analyte is zero at the coating/fluid interface. To ensure that this condition is fulfilled, a high linear flow rate must be generated. The high flow rate is required to ensure that the desorbed analyte is removed immediately from the vicinity of the coating so as not to interact with the coating and slowing the desorption process. Figure 3.13 shows a family of curves which describe changes in the concentration profile in the coating during the desorption process. At the start of the desorption, analyte is removed from the layer of coating close to the interface (Figure 3.13a) and then from the deeper parts of the coating (Figure 3.13b-e). The desorption time profile is presented in Figure 3.14.

As in the perfectly agitated case (Figure 3.5), the desorption process is completed at the time corresponding to $(b - a)^2/2D_f$. The values for the distribution constant and diffusion coefficient will be different compared to the extraction from a perfectly stirred sample because of the different temperature. The above relationship indicate that the desorption time is independent

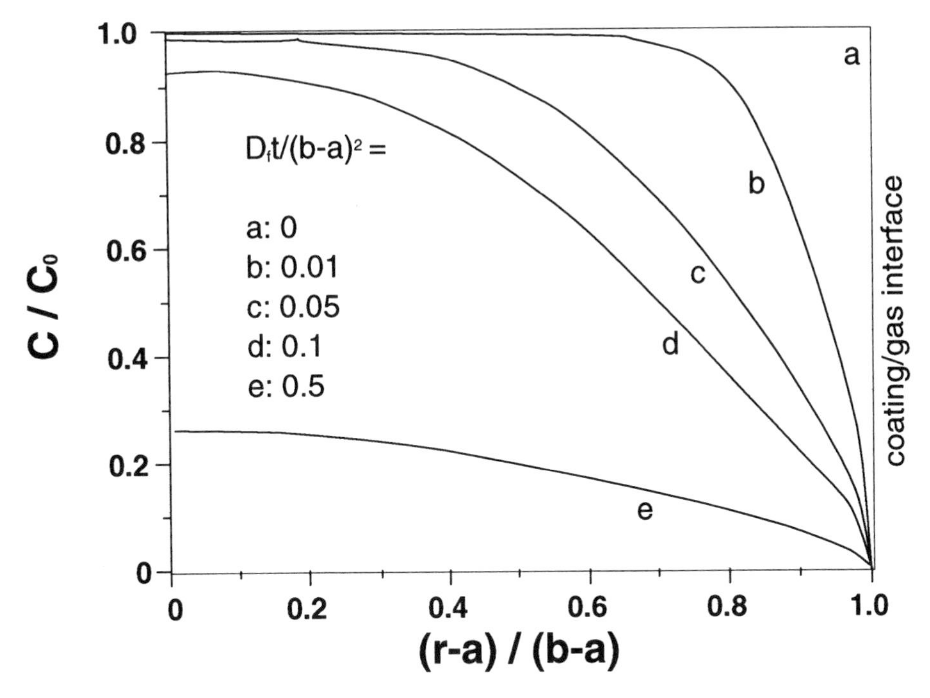

Figure 3.13 Desorption concentration versus radius profile in a fiber coating corresponding to different times after a fiber is exposed in an instrument injection port. The concentration at the coating/carrier-gas interface is assumed to be 0 at all times. The curves have the following values for $Dt/(b - a)^2$: a, 0; b, 0.01; c, 0.05; d, 0.1; e, 0.5.

Source: Adopted with permission from ref. 25.

on the distribution constant K_{fg} and independent of the initial concentration in the fiber. It can be calculated that the desorption time is about 1 second for a 100 μm coating when 200°C is used in the injector for low molecular analytes. In practice however the flow of the mobile phase has finite value and therefore the desorbtion time might be longer for high molecular analytes characterized by the large distribution constants and high molecular weights.

3.3.3 Extraction/Derivatization

Kinetics theory is modified to model extraction into a coating containing a high concentration of reagent which allows simultaneous derivatization and trapping of analytes in the fiber. This in-coating derivatization scheme uses a fiber doped with a reagent, R, that reacts with the analyte, A, to form a chemical product, P, that remains in the fiber:[26]

$$A + R \rightarrow P \tag{3.43}$$

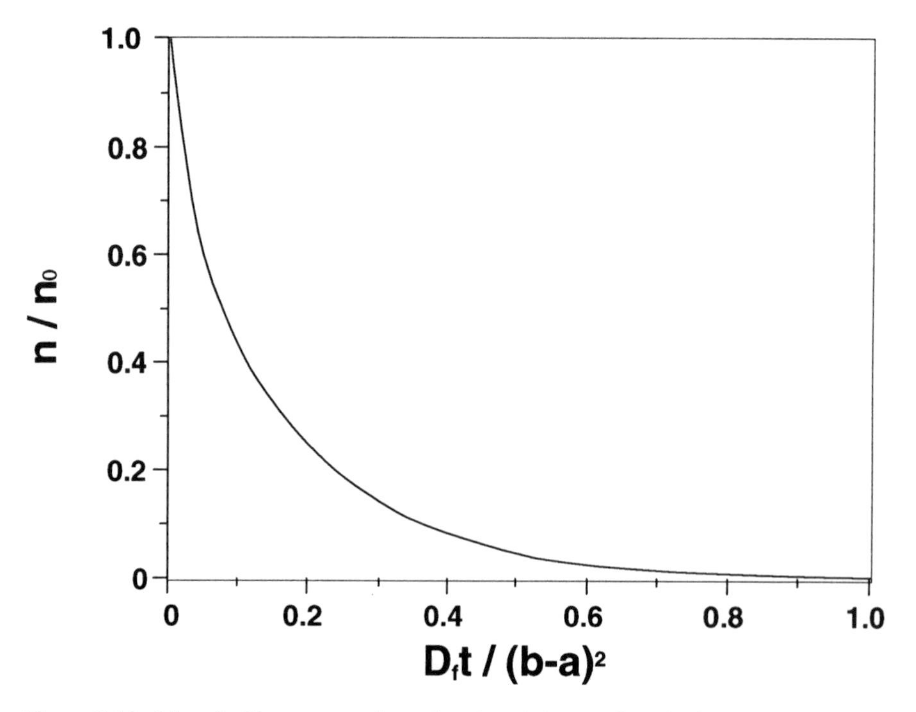

Figure 3.14 Mass in fiber versus time, showing the mass desorbed from a fiber by a fast-moving stripping phase after a fiber is exposed in an instrument injection port.

Source: Adopted with permission from ref. 25.

The reaction rate in the fiber will depend on the concentrations of analyte and reagent in the fiber. The reaction rate is usually described as:

$$\frac{\partial}{\partial t}[\mathrm{P}]_f = k'[\mathrm{R}]_f[\mathrm{A}]_f \tag{3.44}$$

where $[\mathrm{P}]_f$, $[\mathrm{R}]_f$ and $[\mathrm{A}]_f$ represent the concentrations of product, reagent, and analyte in the fiber, respectively and k' is the reaction rate constant.

The theory assumes excess reagent in the fiber, so that reagent concentration remains constant throughout the extraction process, hence reaction rate is proportional only to analyte concentration in the fiber. The term $k'[\mathrm{R}]_f$ in eq. 3.44 is replaced in subsequent equations by the pseudo first-order rate constant k. Under these assumptions, Fick's second law with derivatization in one dimension becomes

$$\frac{\partial[\mathrm{A}]_f}{\partial t} = D\frac{\partial^2[\mathrm{A}]_f}{\partial x^2} - k[\mathrm{A}]_f \tag{3.45}$$

where k is the reaction rate constant. Considering the cylindrical geometry of the fiber and sample system, the equation is converted to:

$$\frac{\partial[\mathrm{A}]_f}{\partial t} = D\frac{1}{r}\left[\frac{\partial}{\partial r}\left(r\frac{\partial[\mathrm{A}]_f}{\partial r}\right)\right] - k[\mathrm{A}]_f \tag{3.46}$$

SPME extraction/derivatization will result theoretically in 100% extraction of the analyte from a vial, assuming that derivatization products remain in the fiber.

The exact analytical solution to this problem is presented in Appendix A. The analytical solution is expressed as a complicated series that is difficult to interpret, so an approximation was derived for the case when KV_f is small compared to V_s. In this case the amount of analyte extracted by the fiber at the partitioning equilibrium, without the reagent present, is insignificant compared to the total amount of analyte in the system. The above condition means that for a 10 mL vial and standard 100 μm thick fiber coating, K_{fs} should be less than 662. This condition is met in most of the cases when derivatization is used since a larger distribution constant is sufficient to provide good sensitivity and therefore the derivatization would be unnecessary.

When the above condition is met, two limiting cases are possible. The first one assumes that the kinetics of the reaction is slow compared to the diffusion of analyte into the fiber, k is small compared to $D_f/(b - a)^2$. In the other words, at any time during the extraction, the fiber coating is at equilibrium with the analyte remaining in a perfectly agitated sample, resulting in uniform reaction rate throughout the coating. This is a typical case, since the equilibration time for well-agitated conditions is very short compared to a typical rate constant. Figure 3.15a illustrates the relationship between the fraction of total amount of analyte accumulated as product in the fiber coating and the exposure time, expressed as dimensionless scale. Initially the reaction rate is defined by:

$$\frac{d[\mathrm{P}]}{dt}(t = 0) = k[\mathrm{A}]_f = kK_{fs}[\mathrm{A}]_s^{t=0} = kK_{fs}C_0 \tag{3.47}$$

where C_0 is the initial concentration of analyte in the sample. In the other words, when the sample is of infinite volume, the reaction and accumulation of analyte in the fiber proceeds with the same speed as long as reagent is present in the excess amount. This accumulation rate at constant concentration conditions is indicated on the graph as a line tangent to the extraction time profile at $t = 0$ (Figure 3.15b). If the concentration varies during the accumulation, the collected amount corresponds to the integral over concetration and time. For limited sample volume, however, the concentration of analyte in the sample phase decreases with time as it is partitioned into the coating and converted to product, resulting in a gradual decrease of the rate. The time required to exhaustively extract analytes in the limited volume case can be estimated from the graph as:

$$t_e = t_{95\%} = \frac{4.6V_s}{kK_{fs}V_f} \tag{3.48}$$

As would be expected, the exhaustive reaction time is proportional to the

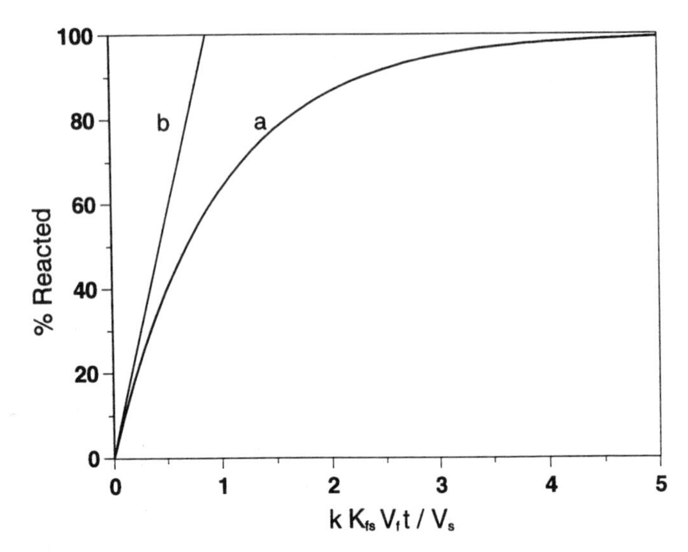

Figure 3.15 (a) Accumulation of reaction products versus time profile for slow reaction in the coating, for SPME "in-fiber" derivatization from a sample of finite volume, as calculated by the mathematical approximation to the complete analytical solution given in Appendix A. This curve is valid when $KV_f < V_s/20$ and $k < D_f/(b - a)^2$. (b) Accumulation of reaction product for infinite volume.

volume of the sample, V_s and inversely proportional to the reaction rate, k, and the capacity of the fiber ($K_{fs}V_f$).

The second limiting case considers the situation when k is large compared to $D_f/(b - a)^2$. In other words, the reaction rate is fast compared to the diffusion rate of analytes into the coating from the perfectly agitated sample. This condition might occur in practice when dense polymer extracting phase are used or the reaction is very fast. The net result is that the product is generated close to the interface between the fiber coating and the sample matrix. Figure 3.16 illustrates the extraction time profile for this case. The time to reach exhaustive removal of analyte from the sample into the coating can be estimated from the Figure 3.16 as:

$$t_e = t_{95\%} = \frac{4.6V_s}{K_{fs}A_f\sqrt{kD_f}} \tag{3.49}$$

As expected based on the above discussion, the time corresponding to exhaustive extraction decreases with increasing fiber surface area, A_f.

3.3.4 Headspace SPME

The geometry of SPME headspace extraction is illustrated in Figure 3.17a.[27] An aqueous sample containing a dissolved organic compound is transferred

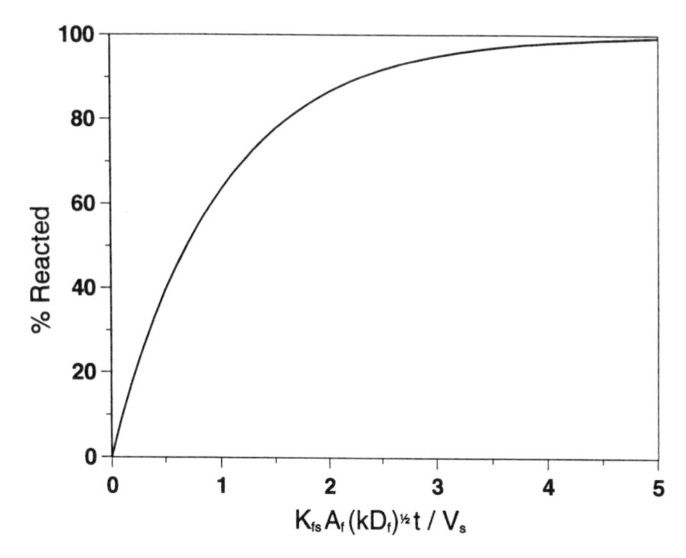

Figure 3.16 Approximate accumulation of reaction products versus time profile for fast reaction in the coating, for SPME "in-fiber" derivatization from a sample of finite volume, as calculated by the mathematical approximation to the complete analytical solution given in Appendix A. This curve is valid when $KV_f < V_s/20$ and $k > D_f/(b - a)^2$.

into a closed container with a headspace. A fused silica fiber coated with a thin layer of a selected liquid organic polymer is inserted into the headspace portion of the container, but does not have any direct contact with the aqueous phase. The fiber's liquid coating starts to absorb the organic analyte from the headspace. The analyte undergoes a series of transport processes, from water to gas phase (or headspace) and eventually to the coating, until the system finally reaches equilibrium. The diffusion process occurs not only in the axial direction, but in the radial direction as well. Exact algebraic solutions of this diffusion problem can be derived. We are more interested in studying the main factors which control the diffusion process, however, rather than the strict mathematical solutions. A simple one-dimensional diffusion model can provide insight into this diffusion problem. In the model illustrated in Figure 3.17b the diffusion occurs in only one direction (x axis), L_f is the thickness of the polymeric coating, L_g is the length of the headspace, L_s is the length of the aqueous solution. The solution to this problem can be found by solving the appropriate mass balance differential equation.

The one-dimensional diffusion process can be described by Fick's second law:

$$\frac{\partial C(x, t)}{\partial t} = D \frac{\partial^2 C(x, t)}{\partial x^2} \tag{3.50}$$

where $C(x, t)$ is the concentration of the analyte at position x and time t, and

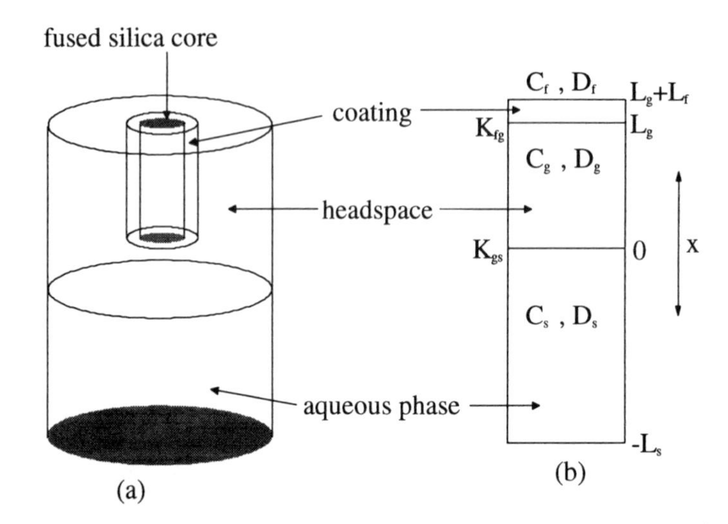

Figure 3.17 (a) Geometry for SPME headspace sampling. (b) One-dimensional model of the three-phase system: K_{fg} and K_{gs} are the coating/headspace and headspace/water partition coefficients; D_f, D_g and D_s are the diffusion coefficients of the analyte in the coating, headspace, and water; C_f, C_h and C_s are the concentrations in the coating, headspace, and water; L_f, L_g and L_s are the thicknesses of coating, headspace, and aqueous phase.

Source: Adopted with permission from ref. 27.

D is the diffusion coefficient of the analyte. The mass of analyte absorbed by the polymeric coating at any given moment, $M(t)$, can be calculated by

$$M(t) = \int_{L_g}^{L_g+L_f} C(x, t)\, dx \tag{3.51}$$

Static Headspace SPME. If we assume that an aqueous solution with initial concentration C_0 is transferred into the container and that the equilibrium between the headspace and the aqueous solution has been reached before headspace SPME, the concentrations in the headspace (C_g^0) and in the aqueous solution (C_s^0) before the extraction can be calculated through $C_0L_s = C_g^0L_g + C_s^0L_s$ with $C_g^0 = K_{gs}C_s^0$. By solving these equations, we can obtain the concentrations of the analyte in any position x at any given moment (refer to Appendix A). Through eq. 3.51, the time profile of the mass absorbed by the coating can also be calculated.

Figure 3.18 shows the concentration profiles from $x = -L_s$ to $x = L_g + L_f$ at different extraction times. The small scale of the x axis has been chosen in order to show clearly the concentration profiles in all three phases. The parameters used here are for the PDMS liquid coating with benzene as the analyte. The diffusion coefficients of benzene are $D_f = 2.8 \times 10^{-6}$ cm^2/s in the coating,[28] $D_g = 0.077$ cm^2/s in the headspace,[29] and $D_s = 1.8 \times 10^{-5}$ cm^2/s in the aqueous solution.[30] A coating thickness $L_f = 56$ μm is used. The

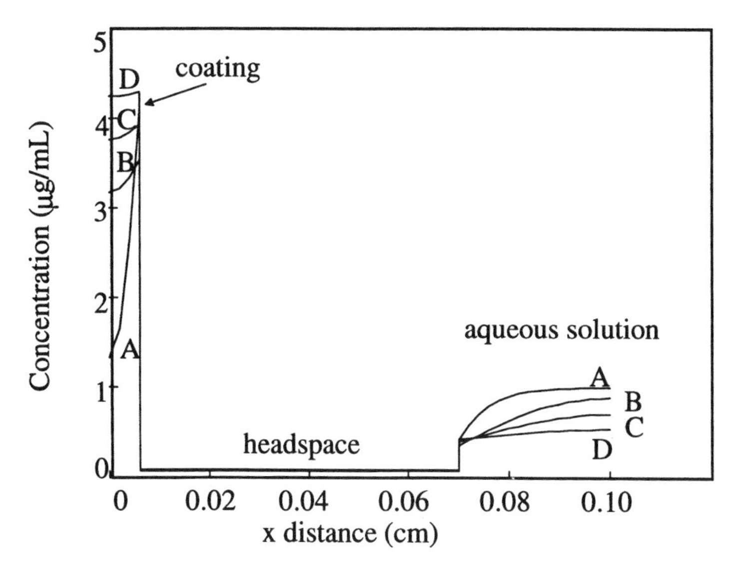

Figure 3.18 Concentration profiles in three phases at different stages of the diffusion process with a static aqueous phase; for: a = 56 μm, b = 0.07 cm, c = 0.1 cm, K_{fg} = 50, K_{gs} = 0.2, $D_f = 2.8 \times 10^{-6}$ cm²/s, D_g = 0.077 cm²/s, $D_s = 1.08 \times 10^{-5}$ cm²/s, and C_0 = 1 μg/mL. The curves have the following values for t: A, 3 s; B, 15 s; C, 30 s; D, 60 s.

thicker the coating, the more analyte is absorbed by the coating, and *vice versa.* Curve A is the concentration profile in all three phases at the beginning of the extraction (t = 3 s). As diffusion continues, the concentration gradient in each phase decreases (from curve B to curve D). The concentrations in both the coating and aqueous phase vary substantially with x in the early stages of the extraction, while the concentration in the headspace is almost constant. This difference is because the diffusion coefficient in the headspace is three or five orders of magnitude higher than in the other two phases.

Equilibrium in the three phases during headspace SPME is reached when the concentration of analyte is homogeneous within each of the three phases and the concentration ratios between two adjoining phases satisfy the distribution constants. The equilibration time, t_e, is set here as the time at which the mass absorbed by the fiber coating has reached 95% of its final total mass, in other words $M(t_e) = 0.95M(\infty)$.

Figure 3.19a illustrates the time profiles of the mass absorbed by the coating for different coating/headspace phase distribution constants, K_{fg}, with a constant headspace/water distribution constant K_{gs} = 0.2. The values of L_f, L_g, and L_s used for this figure are typical experimental values. When K_{fg} becomes large more analytes is absorbed by the coating, and more time is needed to reach equilibrium. For K_{fg} = 100 (curve A), equilibration time t_e is about 47 minutes in comparison with 5.5 hours for K_{fg} = 10,000 (curve C).

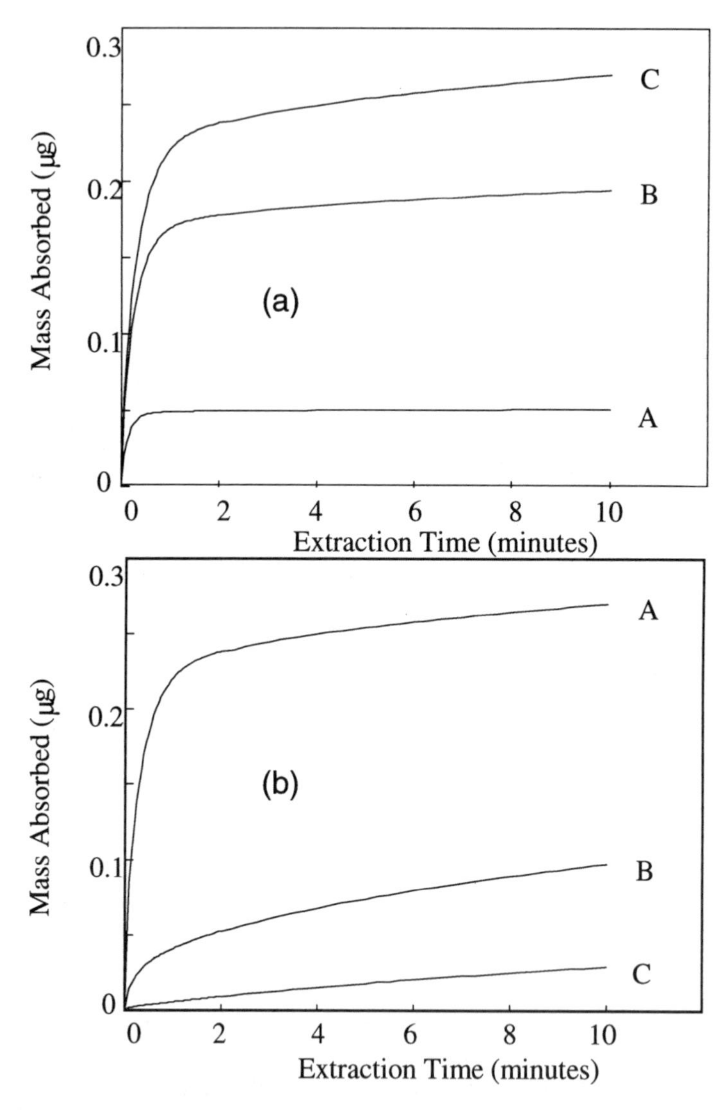

Figure 3.19 (a) Time profiles of the mass absorbed by the fiber coating with a static aqueous phase; the parameters used: $L_f = 56\ \mu\text{m}$, $L_g = 2$ cm, $L_s = 2.5$ cm, $K_{gs} = 0.2$, $D_f = 2.8 \times 10^{-6}\ \text{cm}^2/\text{s}$, $D_g = 0.077\ \text{cm}^2/\text{s}$, $D_s = 1.08 \times 10^{-5}\ \text{cm}^2/\text{s}$, $C_0 = 1\ \mu\text{g/mL}$; (A) $K_{fg} = 100$, (B) $K_{fg} = 1000$, (C) $K_{fg} = 10{,}000$. (b) Time profiles of the mass absorbed by the fiber coating with a static aqueous phase. The values of L_f, L_g, L_s, D_f, D_g, D_s, and C_0 are the same as those in Figure 3.19(a), but $K_{fg} = 10{,}000$; (A) $K_{gs} = 0.002$, (B) $K_{gs} = 0.02$, (C) $K_{gs} = 0.2$.

For $K_{fg} = 100$ (curve A), over 80% of the equilibrium amount $M(\infty)$ is extracted at $t = 40$ s. To reach 95%, 47 minutes is required since the mass transport from the sample matrix is limiting and results in a slowly increasing extraction curve. If K would be smaller than 100 the equilibrium time can be very short since the majority of extraction would be from the headspace only. If K is larger the equilibrium time increases further since a large portion of the extracted analytes originates from the sample matrix. Thus for $K_{fg} = 10{,}000$ (curve C) the equilibration time is 5.5 hours.

The long equilibration time in the case of large K_{fg} is caused by the very slow diffusion of analyte molecules in the aqueous phase. Though the diffusion coefficient in the coating is actually smaller than that in the aqueous phase, the thickness of coating is usually so small compared with the other two phases that the diffusion within the coating takes little time. An interesting phenomenon is that there is a turning point during the extraction process, as shown in Figure 3.19a. The mass absorbed by the coating increases rapidly at the beginning, and then slowly levels off. If we designate t_T as the time needed to reach the turning point, for $K_{gs} = 0.2$ and $K_{fg} = 100$ (curve A), 90% of the total mass has been absorbed by the coating at the turning point, whereas about 50% of the total mass has been absorbed by the coating at t_T when $K_{gs} = 0.2$ and $K_{fg} = 10{,}000$ (curve C). Though the values of K_{fg} change considerably t_T varies little, from about 60 seconds for $K_{gs} = 100$ to about 180 seconds for $K_{gs} = 10{,}000$.

These results suggest that in some cases the mass absorbed by the fiber coating will have reached the defined equilibrium point, 95% of $M(\infty)$, at the turning point. Under these circumstances the headspace extraction method can be used very effectively even with a static aqueous phase. The existence of the turning point is caused by the rapid diffusion process in the headspace followed by very slow diffusion in the aqueous phase. The volume of fiber coating is small, approximately 10^{-4} cm^3. For volatile analytes which have relatively small K_{fg} values and relatively large K_{gs} values (their overall K_{fs} values are small since $K_{fg} \gg K_{gs}$ for most organic compounds) the mass of analyte extracted by the coating is often insignificant compared with the mass of analyte present in the headspace. Thus, the concentration in the aqueous solution is virtually unchanged during extraction. In other words, the extraction time is determined largely by the diffusion in the vapor phase, and the equilibration time is quite short.

Figure 3.19b shows the effect of the headspace/water distribution constant, $K_{hs} = K_{gs}$, on the extraction. Curves A, B, and C show the time profiles of the mass absorbed by the coating with $K_{gs} = 0.002$, 0.02, and 0.2, respectively, and $K_{fg} = 10{,}000$ for all three curves. When the analyte has a small K_{gs} value the concentration of analyte in the headspace is low, the headspace extraction affects the concentration of the aqueous phase, and extraction takes a very long time to reach equilibrium.

The relationships discussed above for specific cases (Figures 3.19 a and b) can be generalized in the following equation for estimating equilibration times:

$$t_e = t_{95\%} = 1.8\left(\frac{L_g}{K_{gw}D_g} + \frac{L_s}{1.6D_s}\right)K_{fw}L_f \tag{3.52}$$

This equation can be used when the amount extracted by the fiber coating is a small portion of analytes present in the sample (see Appendix A). The first term in the parentheses is related to mass transfer in the headspace and the second in the sample. As expected, the mass transfer in the sample is controlled by the diffusion coefficient in the sample, D_s, and the sample length which analytes need to diffuse through, L_s. The rate of the mass transfer in the headspace, on the other hand is related to the gas/sample distribution constant, K_{gw}, diffusion coefficient of analytes in the headspace, D_g, and the dimension of the headspace, L_g. The equilibration time also increases with increase of the fiber capacity, defined as a product of the fiber sample distribution constant, K_{fs}, and the fiber thickness, L_f.

The results in Figure 3.19a,b and Equation 3.52 suggest that headspace SPME has a short equilibration time with a static aqueous phase the analyte should have a large Henry's constant in water, which can occur with high volatility and/or hydrophobicity of the analyte, and should not have a very large distribution constant between the fiber coating and the headspace. Volatile organic compounds usually meet these requirements, thus this method can be used to analyze them very effectively. The extraction as a function of time is illustrated graphically in Figure 3.20 using a dimension-free plot. This plot was calculated for the case when the amount of analyte extracted by the fiber is negligible compared to the amount present in the headspace. This plot

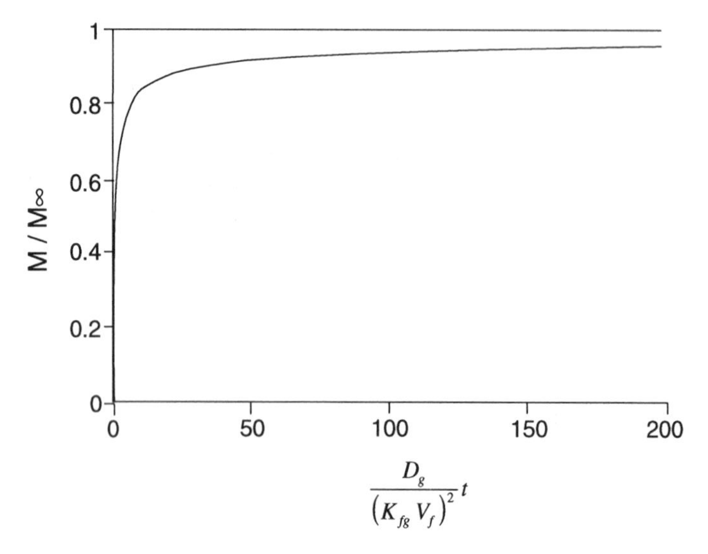

Figure 3.20 Dimensionless plot of time profile for headspace SPME from a large capacity headspace.

should be accurate when $V_g > 12K_{fg}V_f$. The estimated equilibration time when $V_g > 10K_{fg}V_f$ can be estimated by:

$$t_e = t_{95\%} = 100\frac{(K_{fg}V_f)^2}{D_g} \tag{3.53}$$

The above estimate of the equilibration time can be used for all static headspace extractions, including perfect agitation condition in the sample matrix, as long as the extracted analyte originates from the headspace only.

The above discussion suggests that the extraction temperature is a very important parameter to optimize. When the temperature is high enough to move analytes to the headspace from the sample matrix, the extraction occurs only from the headspace, the equilibration time is very rapid and it is independent of the agitation conditions.

Perfect Agitation of The Sample Matrix. In this case, we assume an ideal situation in which the aqueous phase is perfectly mixed, *i.e.* the forced convection in the aqueous solution is infinitely fast and the analyte concentration in the aqueous phase is always homogeneous. This idealization is an accurate representation of reality when the mass transport in the aqueous phase is much faster than that in the other two phases and is not a limiting step for the overall mass transfer process. This can be achieved by stirring the aqueous solution rapidly or using sonication. In the analysis described below the headspace is treated as a static phase throughout the extraction process.

Figure 3.21 shows the concentration profiles in each of the three phases at different extraction times. The parameters used are the same as those for previous figures except the diffusion coefficient of the analyte in the aqueous solution, D_s is no longer used. The main difference between Figures 3.18 and 3.21 is that equilibrium can be achieved more quickly in the latter case. This difference can be seen by comparing curve D in both figures. Also notice that in Figure 3.18, curve D corresponds to 60 seconds while in Figure 3.21 it is 21 seconds. The initial high concentration at the surface of the coating in Figure 3.21A is caused by the rapid diffusion in the headspace and perfect agitation of the aqueous phase. At the beginning of the extraction when the analyte does not have enough time to diffuse deep into the coating, the coating already extracts a substantial quantity of the analyte from the sample and significantly reduces the concentration in the other two phases (compare curve A with the others in the aqueous phase).

Figure 3.22 shows the time profiles of the mass absorbed by the coating for different coating/headspace distribution constants, K_{fg}, with a constant headspace/water distribution constant, $K_{gs} = 0.2$. The larger the K_{fg} value, the longer the equilibration time will be because more analyte molecules need to be transported through the gas phase, which is now the limiting phase. The equilibration time increases from 47 seconds for $K_{fg} = 100$ (curve A) to 220 seconds for $K_{fg} = 10{,}000$ (curve C). These results indicate that with a well-agitated aqueous phase and relatively large K_{gs} values, headspace SPME

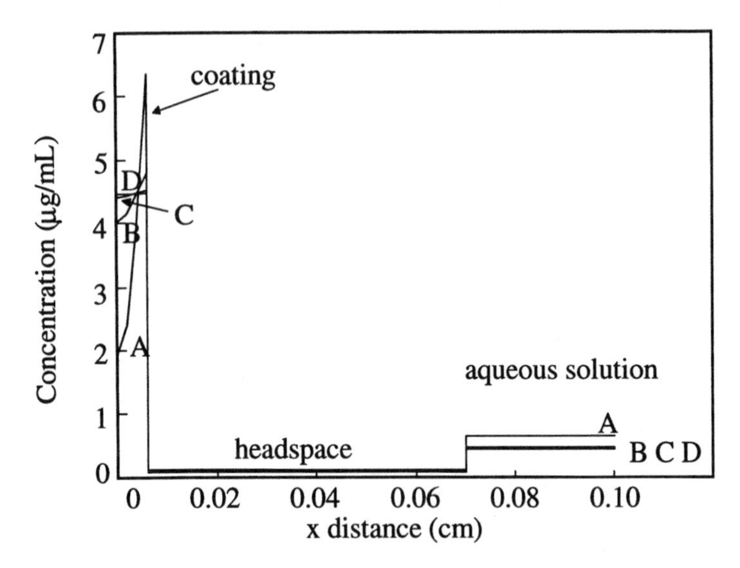

Figure 3.21 Concentration profiles in three phases at different stages of the diffusion process with a well-agitated aqueous phase; the parameters used: L_f = 56 μm, L_g = 0.07 cm, L_s = 0.1 cm, K_{fg} = 50, K_{gs} = 0.2, $D_f = 2.8 \times 10^{-6}$ cm^2/s, D_g = 0.077 cm^2/s, $D_s = 1.08 \times 10^{-5}$ cm^2/s, and C_0 = 1 μg/mL. The curves have the following values for t: A, 3 s; B, 9 s; C, 15 s; D, 21 s.

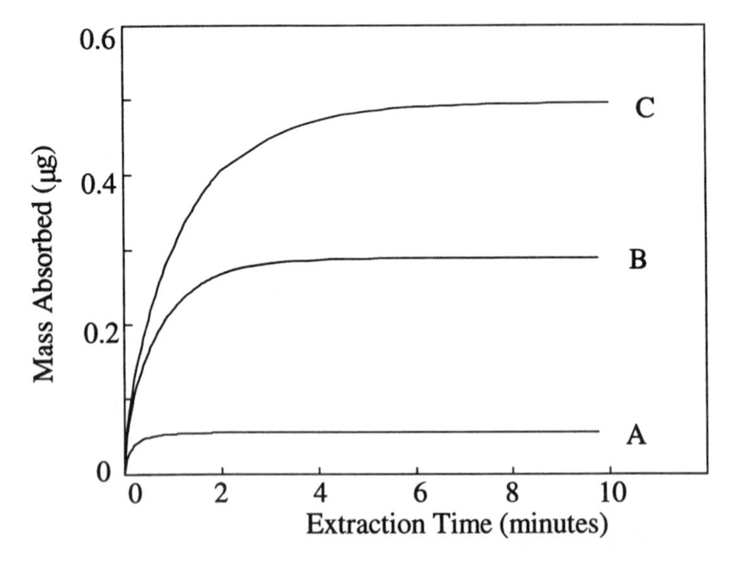

Figure 3.22 Time profiles of the mass absorbed by the fiber coating with an agitated aqueous phase; for: L_f = 56 μm, L_g = 2 cm, L_s = 2.5 cm, K_{gs} = 0.2, $D_f = 2.8 \times 10^{-6}$ cm^2/s, D_g = 0.077 cm^2/s, C_0 = 1 μg/mL; (A) K_{fg} = 100, (B) K_{fg} = 1000, (C) K_{fg} = 10,000.

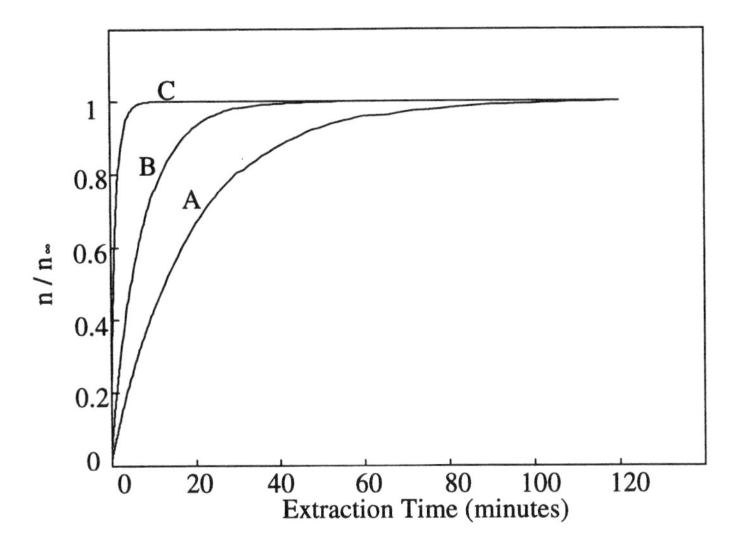

Figure 3.23 Time profiles of the mass absorbed by the fiber coating with an agitated aqueous phase. The values of L_f, L_g, L_s, D_f, D_g, and C_0 are the same as those in Figure 3.22, but K_{fg} = 10,000; (A) K_{gs} = 0.002, (B) K_{gs} = 0.02, (C) K_{gs} = 0.2.

reaches equilibrium in quite a short time even when the distribution constant between the coating and headspace, K_{fg}, is very large.

Figure 3.23 shows the time profiles of the normalized mass absorbed by the coating at different K_{gs} values with the same K_{fg}. The equilibration time becomes longer as the K_{gs} values (or Henry's constants) of organic compounds become smaller. For K_{gs} = 0.2 and K_{fg} = 10,000 (curve C), the equilibration time is 220 seconds, while for K_{gs} = 0.002 and K_{fg} = 10,000 (curve A), the equilibration time is about 60 minutes. Since the aqueous phase is well agitated and the coating is very thin, the limiting step is the diffusion in the headspace from the headspace/water interface to the coating/headspace interface. As Fick's first law points out, the rate of diffusion depends on both the diffusion coefficient and the concentration gradient:

$$J = -D\frac{\partial C}{\partial x} \tag{3.54}$$

where J is the flux of the analyte. In the headspace, the diffusion coefficients of analytes are relatively large, but the concentration gradients become smaller and smaller as Henry's constants decrease if the length of the headspace is unchanged. As a result, the transport of analyte molecules with low Henry's constants through the headspace is very slow and equilibrium takes a long time to achieve.

Figure 3.24 shows the effect of K_{gs} on the equilibration time. For all points in the figure the distribution constant between the coating and water, K_{fs} = $K_{fg}K_{fg}$, is the same so the larger the K_{fg} is, the smaller the K_{gs} will be, and *vice*

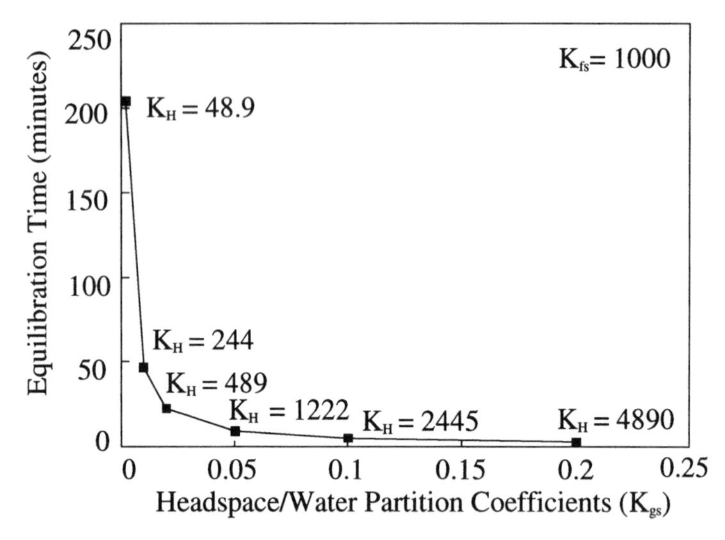

Figure 3.24 Equilibration time versus K_{gs} with $K_{fs} = 1000$; the aqueous phase is well agitated. The points in the figure are as follows: $K_{gs} = 0.2$, 0.1, 0.05, 0.02, 0.01, and 0.002; the corresponding K_H values (shown in units of atm · cm^3/mol).

versa. In the figure the corresponding Henry's constants in units of atm · cm^3/mol, K_H, are also shown. Thus for $K_{fs} = 1000$ when $K_{gs} = 0.002$, which corresponds to Henry's constant $K_H = 48.9$ atm · cm^3/mol at $T = 298$ K, then $K_{fg} = 5 \times 10^6$ and the equilibration time could be more than 3 hours, even with a well-agitated aqueous phase. Because this is a simplified model, Figure 3.24 cannot be used directly to determine the equilibration time for an analyte in a real experiment, but the figure does show the general relationship between the equilibration time and the headspace/water distribution constant (K_{gs}), or Henry's constant (K_H).

For compounds that have a coating/water distribution constant, ($K_{fs} \ll 1000$), the equilibration time does not change significantly as the headspace/water distribution constant, K_{gs}, becomes smaller. In Figure 3.25, $K_{fs} = 20$, 2, and 0.2 respectively as K_{gs} decreases from 0.2 to 0.02 to 0.002 (curves A, B, and C) when the coating/headspace distribution constant, K_{fg}, is kept at 100. Equilibration time in these three cases differ very little. This small change is because the mass of analyte absorbed by the coating is mostly from the headspace when K_{fs} values are small. There is no need for a large quantity of the analyte to be transported from the aqueous phase through the headspace to the coating. Consequently, the slow transport rate of less volatile compounds in the headspace does not have a major impact on the equilibration time when K_{fs} is small. As K_{fs} values become smaller, however, the mass absorbed by the fiber coating becomes smaller, which leads to lower sensitivity. For very volatile compounds (K_{gs} is large), small K_{fs} values mean small K_{fg} values. If

K_{fg} values are about 1, it means that the fiber coating does not have any concentrating effect. In that case, headspace syringe injection will be preferable, since it can inject a larger volume onto a GC column.

Practical Agitation Conditions. The discussion above relates to a theoretical situation when the sample matrix is perfectly aggitated and the headspace is perfectly static. In practice, perfect agitation of the matrix will cause a similar condition in the headspace. In the case when perfect agitation conditions are in both phases, equation 3.35 can be used to estimate equilibration times. For more realistic agitation conditions, modified equations 3.52 and 3.53 can be used to estimate the equilibration times. A theoretical approach is to compute an eddy diffusion coefficient, D_e, which describes the agitation effect on mass transport. This eddy diffusion coefficient is added to the molecular diffusion coefficient to obtain an approximate "virtual" diffusion coefficient.[31] For example, Fick's first law could be adapted to describe mass flux in a stirred water sample as follows:

$$J = (D_e + D_s)\frac{\partial C}{\partial x}. \qquad (3.55)$$

A relation to calculate the eddy diffusion coefficient is

$$D_e = 0.0075\nu R_d \qquad (3.56)$$

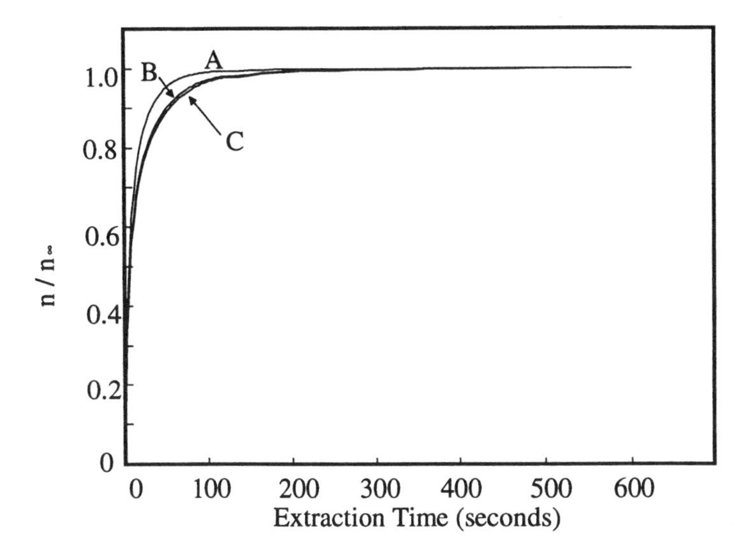

Figure 3.25 Time profiles of the mass absorbed by a fiber coating with an agitated aqueous phase. The values of L_f, L_g, L_s, D_f, D_g, and C_0 are the same as those in Figure 3.22, but $K_{fg} = 100$; (A) $K_{gs} = 0.2$, (B) $K_{gs} = 0.02$, (C) $K_{gs} = 0.002$.

where ν is the kinematic viscosity and R_d the Reynolds number of the fluid.[32] For fluid stirred in a vial the rotating Reynolds number is given by

$$R_d = 4R^2N\rho/\nu \tag{3.57}$$

where R is the impeller radius, N is rate of rotation, ρ is the fluid's density.[33] Headspace of magnetically stirred sample could be considered fluid in a cylindrical housing with a rotating disk at the bottom. Based on eqs. 3.41 and 3.42 the average rotational speed of the water surface would be

$$N_w = \frac{1}{R}\int_0^R \frac{u(r)}{2\pi r}\,dr = 0.55N \tag{3.58}$$

assuming stir bar radius is close to vial radius. This relationship can be used in eq. 3.57 to calculate the headspace Reynolds number.

For example, for headspace SPME from magnetically stirred water in a vial equation eq. 3.52 becomes

$$\begin{aligned} t_e &= t_{95\%} \\ &= 1.8\left(\frac{L_g}{K_{gw}(D_g + 2\cdot 10^{-5}NR^2)} + \frac{L_s}{1.6(D_s + 0.03\cdot NR^2)}\right)K_{fw}L_f \end{aligned} \tag{3.59}$$

where R is the radius of the vial and N is the revolution rate of the stir bar, assumed to have radius close to that of the vial. Other equations with diffusion coefficients as parameters could be modified for agitated systems in a similar way.

3.3.5 In-tube SPME

The discussion above is limited to systems involving extraction phases dispersed on the surface of a rod which is exposed from the protective tubing (needle). In addition to this approach, in some solid phase microextraction systems discussed in Chapter 2, the extraction phase remains in the tubing during the extraction. These approaches involve either coated rods retracted into the needles (Figure 2.28A), or an extracting phase is dispersed onto the inner surface of the tubing (Figure 2.2B and 2.28B). Below we briefly discuss the theoretical aspects of the extraction process which use these geometric arrangements.

Dynamic In-tube SPME. Assume that the system consists of a piece of fused silica capillary, internally coated with a thin film of extracting phase (a piece of open tubular capillary GC column), or that the capillary is packed with extracting phase dispersed on an inert supporting material (a piece of micro-LC capillary column). In these geometric arangements, the concentration profile along axis, x, of the tubing containing the extracting phase as a function of time, t, can be described by adopting the expression for dispersion of the concentration front[34,35]:

$$C(x,t) = \frac{1}{2} C_0 \left[1 - erf \frac{x - \dfrac{ut}{1 + k_p}}{\sigma \sqrt{2}} \right] \tag{3.60}$$

where u is linear velocity of the fluid through the tube, k_p is the partition ratio defined as:

$$k_p = K_{fs} \frac{V_f}{V_V} \tag{3.61}$$

where K_{fs} is a sample/coating distribution constant, V_f is the volume of the extracting phase and V_V is a void volume of the tubing containg the extracting phase. σ is the mean square root dispersion of the front defined as:

$$\sigma = \sqrt{Ht \frac{u}{1 + k_p}} \tag{3.62}$$

where H is equivalent to the HETP (height equivalent to theoretical plate) in chromatographic systems—this can be calculated as a sum of individual contributions to the front dispersion. These contributions are dependent on the particular geometry of the extracting system[35].

Equation 3.60 indicates that the front of analyte migrates through the capillary with a speed proportional to the linear velocity of the sample, and inversely related to the partition ratio. For short capillaries with a small dispersion, the extraction time can be assumed to be similar to the time required for the centre of the band to reach the end of the capillary:

$$t_e = \frac{L \left(1 + K_{fs} \dfrac{V_f}{V_V} \right)}{u} \tag{3.63}$$

where L is the length of the capillary holding the extraction phase. As expected, the extraction time is proportional to the length of the capillary and inversely proportional to the linear flowrate of the fluid. Extraction time also increases with an increase in the coating/sample distribution constant and with the volume of the extracting phase but decreases with an increase in the void volume of the capillary. It should be emphasized that above equation can be used only for direct extraction when the sample matrix passes through the capillary. This approach is limited to particulate-free gas and clean water samples. The headspace SPME approach can broaden the application of the in-tube SPME. In that case, careful cosideration to the mass transfer between sample and headspace should be given in order to describe the process properly (analogous to the discussion in 3.3.4). Also, if the flowrate is very rapid producing turbulent behaviour and the coating/sample distribution constant is not very high, then the perfect agitation conditions are met and the equation 3.35 can be used to estimate equilibration times.

Removal of analytes from a tube is an elution problem analogues to frontal chromatography and has been discussed in detail in reference 35. In general, if the desorption temperature of a GC is high and thin coatings are used, then all the analytes are in the gas phase as soon as the coating is placed in the injector, and the desorbtion time corresponds to the elution of two void volumes of the capillary. For liquid desorption (for example into LC system, Figure 2.28B), the desorbtion volume can be even smaller since the analytes can be focused at the front of the desorption solvent.[36]

Static In-tube SPME Sampling. In addition to the analyte concentration measurement at a well defined place in space and time, obtained by using the approaches discussed above, an integrating sampling is possible with a simple SPME system. This is particularly important in field measurements when changes of analyte concentration over time, and place to place variations, must often be taken into account.

When the extracting phase is not exposed directly to the sample, but is contained in the protective tubing (needle) without any flow of the sample through it (as discussed above), the extraction occurs through the static gas phase present in the needle. The integrating system can consist of extracting phase coating the interior of the tubing as shown in the Figure 2.2B, or it can be an externally coated fibre withdrawn into the needle as shown in Figure 2.28A. These geometric arangements represents a very powerful method able to generate a response proportional to the integral of the analyte concentration over time and space (when the needle is moved through the space)[37]. In these cases, the only mechanism of analyte transport to the extracting phase is diffusion through the gaseous phase contained in the tubing. During this process, a linear concentration profile is established in the tubing between the small needle opening, characterized by surface area A and the position of the extracting phase, located at the distance Z from the opening. The amount of analyte extracted, dn, during time interval, dt, can be calculated by considering the first Fick's law of diffusion[38]:

$$dn = AD_g \frac{dc}{dz} dt = AD_g \frac{\Delta C(t)}{Z} dt \tag{3.64}$$

where $\Delta C(t)/Z$ is a value of the gradient established in the needle between needle opening and the position of the extracting phase, Z; $\Delta C(t) = C(t) - C_Z$, where C(t) is a time dependent concentration of analyte in the sample in the vicinity of the needle opening, and C_Z concentration of the analyte in the gas phase in the vicinity of the coating. C_Z is close to zero for a high coating/gas distribution constant capacity, then: $\Delta C(t) = C(t)$. The concentration of analyte at the coating position in the needle, C_Z, will increase with integration time, but it will be kept low compared to the sample concentration in the sample $C(t)$ because of the presence of the sorbing coating. Therefore the accumulated amount over time can be calculated as:

$$n = D_g \frac{A}{Z} \int C(t)\, dt \quad (3.65)$$

As expected, the extracted amount of analyte is proportional to the integral of a sample concentration over time, the diffusion coefficient of analytes in gaseous phase, D_g, area of the needle opening, A, and inversely proportional to the distance of the coating position in respect of the needle opening, Z. It should be emphasized that equation 3.65 is valid only in a situation where the amount of analyte extracted onto the sorbent is a small fraction (below RSD of the measurement, typically 5%) of equilibrium amount in respect to the lowest concentration in the sample. To extend integration times, the coating can be placed deeper into the needle (larger Z), the opening of the needle can be reduced by placing an additional orifice (smaller A), or a high capacity sorbent can be used. The first two solutions will result in a low measurement sensitivity. An increase of sorbent capacity presents a more atractive opportunity. It can be achieved by either increasing the volume of the coating, or its affinity towards the analyte. An increase of the coating volume will require an increase of the device size. The optimum approach to increased integration time, is to use sorbents characterized by large coating/gas distribution constants.

3.4 Experimental Verification

Some of the useful conclusions given in this chapter on SPME theory have been verified experimentally and are discussed below:

1. **The concentration of the sample has no impact on the concentration time profile and the equilibration time**. Figure 3.26 demonstrate this fact experimentally. In other words, if the extraction is optimized for a given concentration, the equilibration time will be the same for other concentrations as well. This condition is valid as long as the system behaves linearly, in the other words, distribution constants between the various components in the SPME/sample system remain constant with a concentration change.
2. **The agitation condition determines the extraction rate and equilibration time for extraction from aqueous samples.** Figure 3.27 shows experimental data comparing equilibration time profiles obtained for magnetic stirring conditions at various rotational speeds. Figure 3.28 illustrates that increased mass transfer conditions can be obtained by moving the fiber with respect to solution (Figure 3.28C) and by vibrating the vial in a sonicator bath (Figure 3.28B). In headspace SPME, agitation allows faster extraction of less volatile species. Figure 3.29 indicates similar extraction times for benzene under static or agitated conditions, but the equilibration time profiles are already substantially different for o-xylene. Figure 3.30 shows that sonication is more effective way of facilitating mass transfer of analytes from the sample matrix to the fiber compared to stirring.

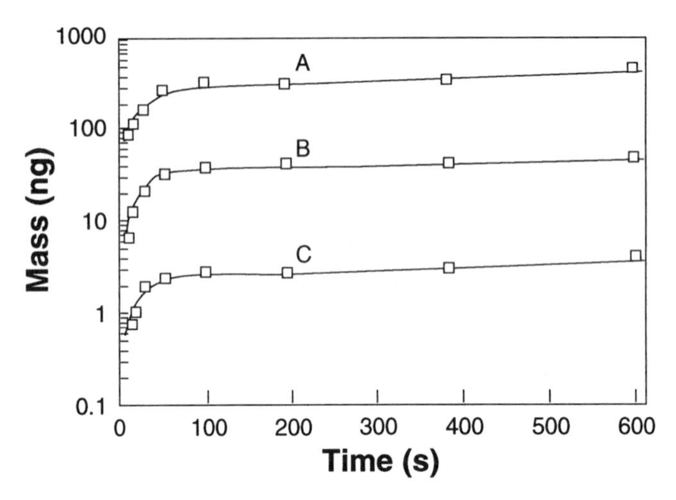

Figure 3.26 Effect of analyte concentration on the absorption vs time profile of 2500 rpm stirred benzene in water extracted with a 56 μm thick coating on a 1 cm long fiber, for: $K_{fs} = 125$, $a = 0.007$ cm, $b = 0.0126$ cm, $L = 1$ cm, $D_s = 1.08 \times 10^{-5}$ cm^2/s, $D_f = 2.8 \times 10^{-6}$ cm^2/s. (A) $C_s = 10$ ppm, (B) $C_s = 1$ ppm, (C) $C_s = 0.1$ ppm.

Source: Adopted with permission from ref. 25.

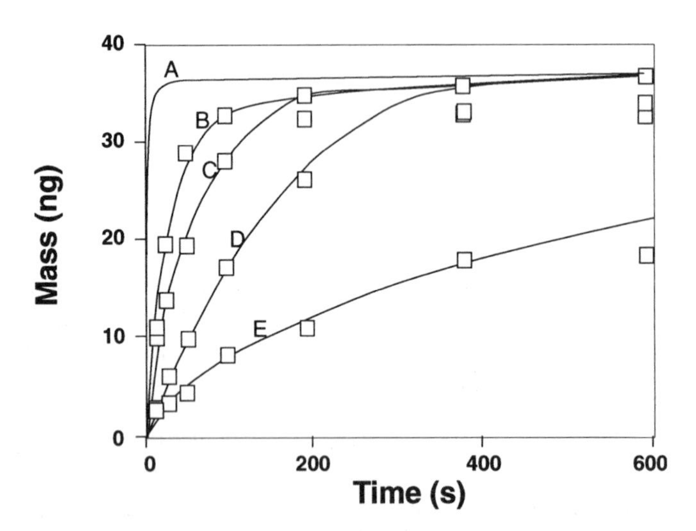

Figure 3.27 Effect of stirring on the absorption versus time profile for the extraction of 1 ppm benzene in water. Parameters are the same as given for Figure 3.26. Stir bar 7 mm long, fiber positioned at center of vial, 7.4 mL vial. Curves are as follows: (A) theoretical prediction for perfect agitation; (B) 2500 rpm; (C) 1800 rpm; (D) 400 rpm; (E) no stirring.

Source: Adopted with permission from ref. 25.

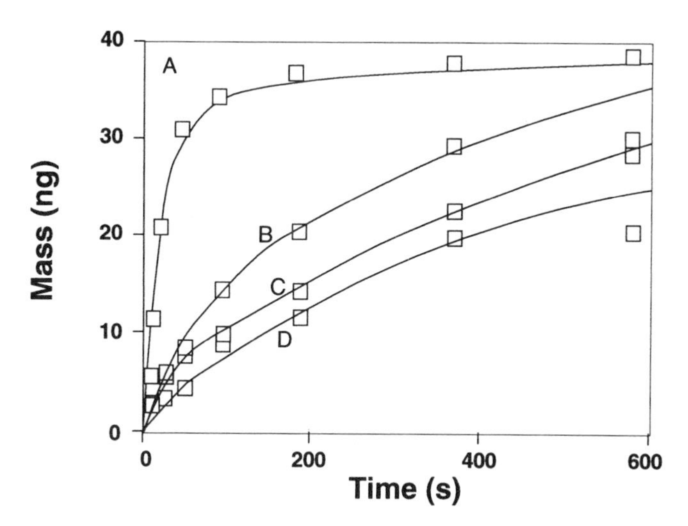

Figure 3.28 Effect of agitation method on the absorption versus time profile for extraction of 1 ppm benzene in water for parameters as given for Figures 3.26 and 3.27. Curves are as follows: (A) 2500 rpm magnetic stirring; (B) sonication with 1/8 in. horn disrupter tip placed in 50 mL vial at a low power less than 100 W; (C) manually repeated fiber insertion/retraction, at a rate of one insertion and retraction per second; (D) no stirring.

Source Adopted with permission from ref. 25.

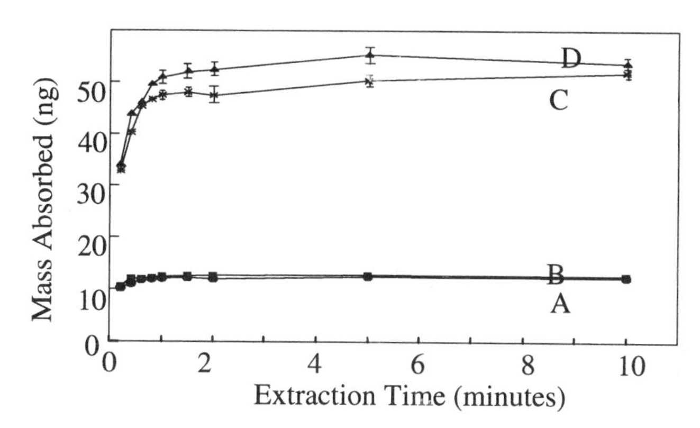

Figure 3.29 The extraction time profiles of 1 ppm benzene and *o*-xylene in aqueous solutions during headspace SPME: (A) benzene with a static aqueous phase; (B) benzene with a well-agitated aqueous phase; (C) *o*-xylene with a static aqueous phase; (D) *o*-xylene with a well-agitated aqueous phase.

Source: Adopted with permission from ref. 27.

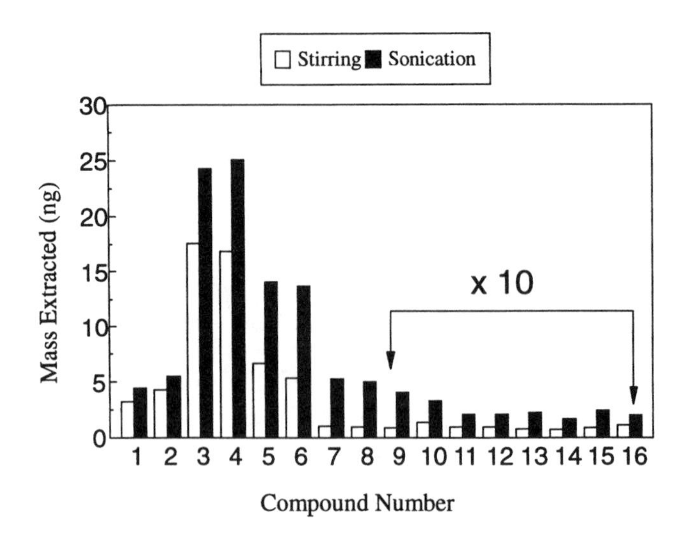

Figure 3.30 The mass of PAHs extracted by the fiber coating from aqueous samples of 0.1 ppm PAHs using headspace SPME under either sonication or magnetic stirring; bars from compounds 9 to 16 were enlarged 10 times. The compounds are: 1, naphthalene; 2, acenaphthylene; 3, acenaphthene; 4, fluorene; 5, phenanthrene; 6, anthracene; 7, fluoranthene; 8, pyrene; 9, benzo[a]anthracene; 10, chrysene; 11, benzo[b]fluoranthene; 12, benzo[k]fluoranthene; 13, benzo[a]pyrene; 14, indo[1,2,3-cd]pyrene; 15, dibenzo[a,h]anthracene; 16, benzo[ghi]perylene.

3. **Extraction time is affected by coating thickness.** Figure 3.31 shows experimental results clearly demonstrating that coating thickness changes not only the amount of analyte extracted, but also the equilibration time. It is important to use the thinnest coating which gives acceptable sensitivity.
4. **The distribution constant affects the equilibration time.** Figure 3.32 shows experimental data for BTEX and differences in the corresponding extraction time profiles. The amount of analyte extracted increases with K_{fs}, but the equilibration time becomes longer as well. The compound which has higher affinity toward the coating reaches equilibrium later . Using a thin selective coating will not help the extraction times for aqueous and other heterogeneous matrices, since in these cases the transport to the coating controls the rate of extraction.
5. **Temperature is a very important parameter to optimize.** An increase in extraction temperature translates to an increased diffusion coefficients and decreased distribution constant, both leading to faster equilibration time. This effect can be applied to optimize the extraction time for a given coating. If sensitivity is high enough at a higher temperature, increasing the temperature can lead to faster determinations. In general, the highest possible temperature should be used. In headspace SPME, an increase in extraction temperature also leads to an increase of analyte concentration in the headspace, and helps to facilitate faster extraction. Figure 3.33 illus-

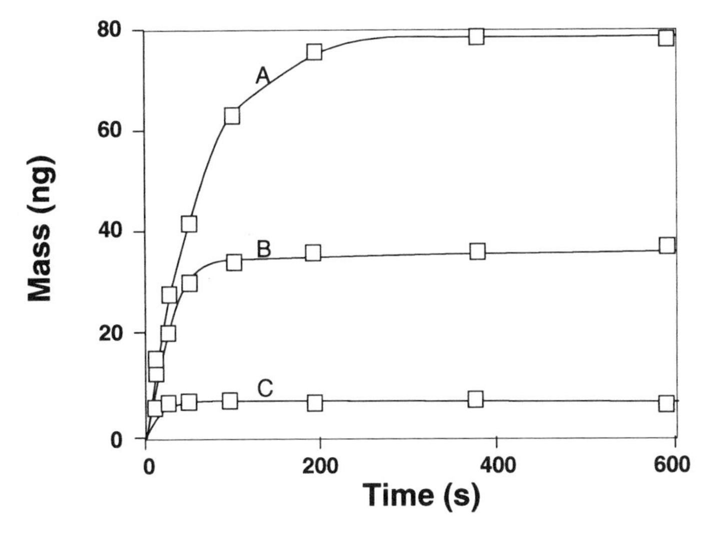

Figure 3.31 Effect of coating thickness on the absorption versus time profile for the extraction of 0.1 ppm benzene from a 2500 rpm stirred aqueous solution. Parameters as given for Figures 3.26 and 3.27, except for fiber coating inner and outer radius: (A) 100 μm thick coating, a = 0.0055 cm, b = 0.0155 cm; (B) 56 μm thick coating, a = 0.007 cm, b = 0.0126 cm; (C) 15 μm thick coating, a = 0.0055 cm, b = 0.0070 cm.

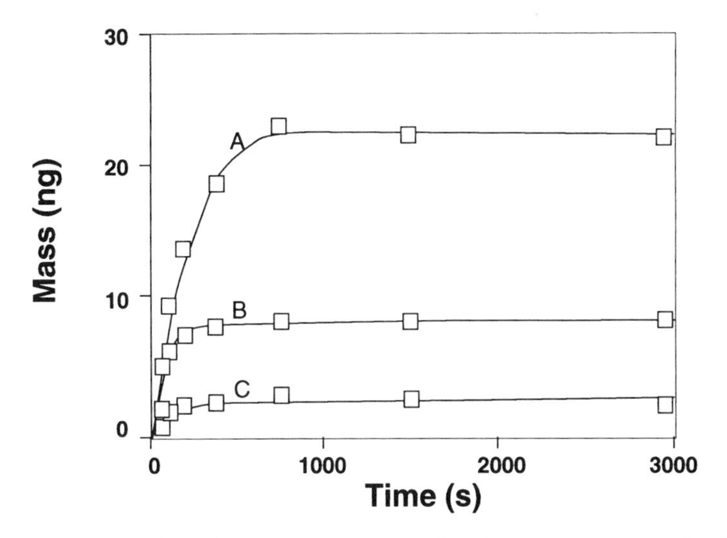

Figure 3.32 Effect of distribution constant on the absorption profile of 2500 rpm stirred 0.1 ppm analyte extracted with a 56 μm thick coating on a 1 cm long fiber. Parameters as given for Figures 3.26 and 3.27, except for distribution constants. The curves represent p-xylene, K_{fs} = 831 (A), toluene, K_{fs} = 294 (B), and benzene, K_{fs} = 125 (C).

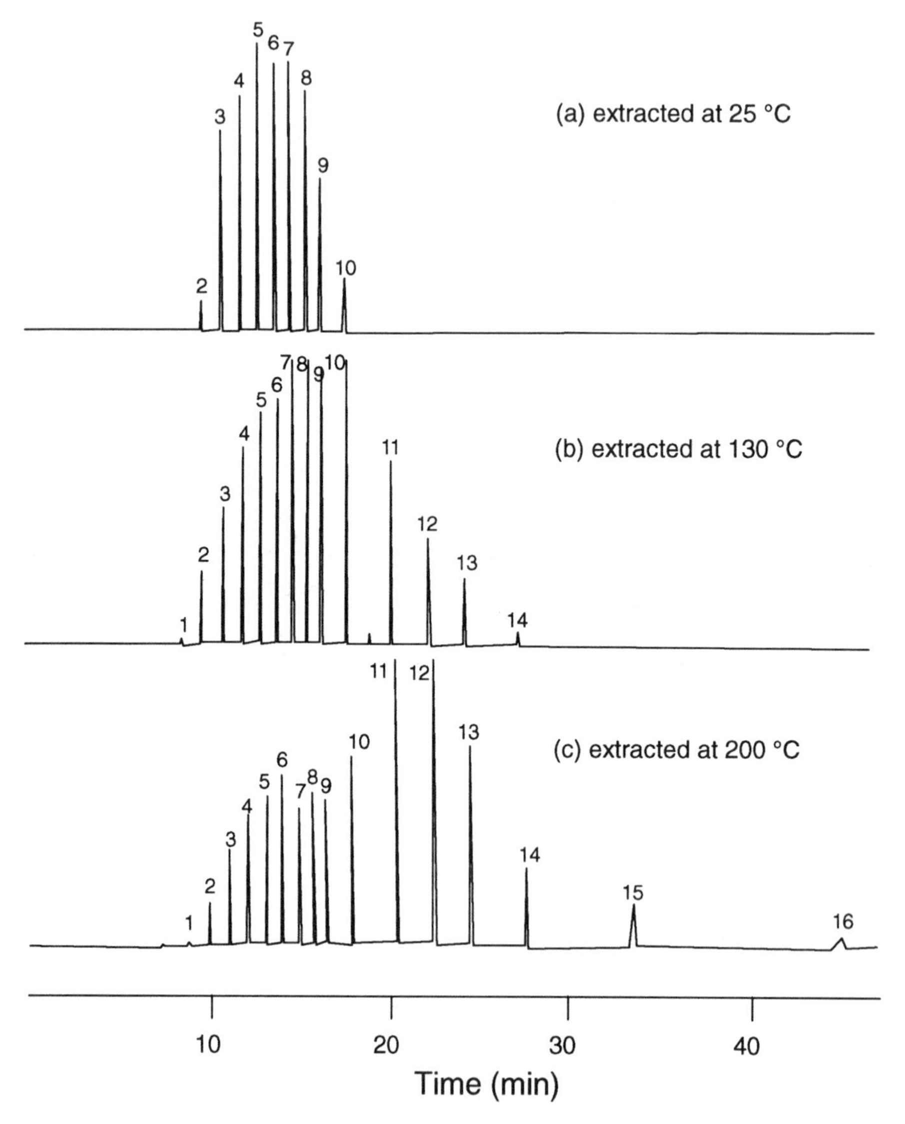

Figure 3.33 Total ion current chromatogram of 16 straight chain hydrocarbons sampled by Headspace SPME from spiked sand at 25°C (a), 130°C (b) and 200°C for 60 minutes: 1, C10; 2, C11; 3, C12; 4, C13; 5, C14; 6, C15; 7, C16, 8, C17; 9, C18; 10, C20; 11, C24; 12, C28; 13, C32; 14, C36; 15, C40.

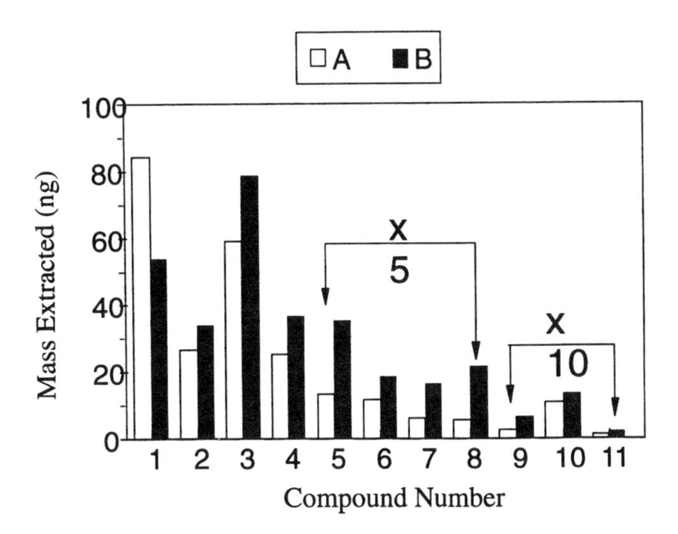

Figure 3.34 The mass extracted from sand samples spiked with 0.02 ppm PAHs and 5% water during headspace SPME; A, 2-minute sampling at 100 °C; B, microwave heating (600 W) for 80 seconds. The bars from 5 to 8 multiplied by 5 and from 9 to 11 by 10. The compound names are listed in the caption for Figure 3.30.

trates that increased extraction temperature results in increase of extraction rate for high molecular weight compounds. However the increase of temperature results in a decrease of the K_{fs} and loss of sensitivity for less volatile species. Figure 3.34 illustrates that rapid microwave heating releases analytes from the matrix more effectively than conventional heating.[39]

References

1. P. W. Atkins. *Physical Chemistry, 4th* Edition. (Freeman, New York, 1978.)
2. J. Langenfeld, S. Hawthorne, D. Miller, J. Pawliszyn, *Anal. Chem.* ***67***, 1727 (1995).
3. N. Alexandrou, J. Pawliszyn, *Anal. Chem.* ***61***, 2770 (1989).
4. Z. Zhang, J. Pawliszyn, *J. Phys. Chem.* ***100***, 17648 (1996).
5. P. Martos, A. Saraullo, J. Pawliszyn, *Anal. Chem.* ***69***, 402 (1997).
6. R. Schwarzenbach, P. Gschwend, D. Imboden. *Environmental Organic Chemistry.* (John Wiley & Sons Inc., New York, 1993), pp. 109–123.
7. J. Li, P. W. Carr, *J. Chromatogr. A* ***659***, 367 (1994).
8. J. H. Park, J. E. Lee, M. D. Jang, J. Li, P. W. Carr, *J. Chromatogr.* ***586***, 1 (1991).
9. A. J. Dallas, P. W. Carr, *J. Phys. Chem.* ***98***, 4927 (1994).
10. C. L. Arthur, L. M. Killam, K. D. Buchholz, J. Pawliszyn, *Anal. Chem.* ***64***, 1960 (1992).
11. P. Martos, J. Pawliszyn, *Anal. Chem.* ***69***, 206 (1997).

12. F. A. Long, W.F. McDevit, *Chem. Rev.* ***51***, 119 (1952).

13. X. Yang, T. Peppard, *LC-GC* ***13***, 882 (1995).

14. J. P. Conzen, J. Burck, H. J. Ache, *Applied Spec.* ***47***, 753 (1993).

15. J. Brandrup, E. H. Immergut, Eds. *Polymer Handbook, 3rd Ed.* (John Wiley & Sons, Toronto, 1989.)

16. *Handbook of Chemistry and Physics, 70th Ed.* (CRC Press, Boca Raton, 1989.)

17. G. L. Klunder, R. E. Russo, *Appl. Spec.* ***49***, 379 (1995).

18. Z. Y. Zhang, J. Pawliszyn, *Anal. Chem.* ***67***, 34 (1995).

19. The Thermodynamic Research Center. *TRC Thermodynamic Tables.* (The Texas A&M University System, College Station, Texas, 1992.)

20. H. S. Carslaw, J .C. Jaeger. *Conduction of Heat in Solids, 2nd Edition.* (Clarendon Press, Oxford, 1986.)

21. A. D. Young. *Boundary Layers.* (BSP Professional Books, Oxford, 1989.)

22. The mathematics of diffusion and heat transfer are equivalent because both processes are described by Laplace's equation. Formulae for heat transfer can be used for mass transfer by substituting diffusion coefficients for thermal conductivity and diffusivity constants (see reference [20], section 1.13.I). The term D/d in equation (3.39) corresponds to the heat transfer coefficient in Newton's formula for heat transfer at interfaces. Therefore the formula for heat transfer at interfaces can be used to estimate d, the thickness of an effective diffusion boundary layer.

23. A. Mikheyev. *Fundamentals of Heat Transfer.* (Mir Publishers, Moscow, 1966.)

24. A. Ogawa. *Vortex Flow.* (CRC Press, Boca Raton, Florida, 1993.)

25. D. Louch, S. Motlagh, J. Pawliszyn, *Anal. Chem.* ***64***, 1552 (1992).

26. L. Pan, M. Adams, J. Pawliszyn, *Anal. Chem.* ***67***, 4396 (1995).

27. Z. Zhang, J. Pawliszyn, *Anal. Chem.* ***65***, 1843 (1993).

28. A. C. Newns, G. S. Park, *J. Polym. Sci., Part C* **22**, 927 (1969).

29. E. W. Washburn. *International Critical Tables of Numerical Data, Physics, Chemistry and Technology.* (McGraw-Hill, New York, 1926), Vol. 5, p. 62.

30. J. O. L. Wendt, G. C. Frazler, Jr. *Ind. Eng. Chem. Fundam.* ***12***, 239 (1973).

31. D. K. Edwards, V. E. Denny, A. F. Mills. *Transfer Processes,* 2nd edition. (Hemisphere Publishing Corp., New York, 1979.)

32. G. T. Orlob. *Contribution no. 19: Eddy Diffusion in Open Channel Flow.* (Water Resources Center, University of California, Berkely, California, 1958.)

33. *Encyclopedia of Science and Technology,* 7th edition. (McGraw-Hill, Toronto, 1992.)

34. J. Crank. *Mathematics of Diffusion.* (Clarendon Press, Oxford, 1989.), p 14.

35. J. Pawliszyn, *J. Chromatogr. Sci.* ***31***, 31 (1993).

36. R. Eisert and J. Pawliszyn, *Anal. Chem., submitted.*

37. M. Chai and J. Pawliszyn *Environ. Sci. Technol.* ***29***, 693 (1995).

38. J. Crank. *Mathematics of Diffusion.* (Clarendon Press, Oxford, 1989.), p 2.

39. Z. Zhang. *Ph.D. Thesis: Headspace Solid Phase Microextraction.* (University of Waterloo, Waterloo, Canada, 1995.)

CHAPTER

4

SPME Method Development

This chapter discusses the stages in SPME method development and the experimental experiences gained with this technique. The stages are summarized in Table 4.1. In most cases, not all steps have to be completed, particularly the initial ones, since knowledge gained from previous experiments as well as from literature, can be applied to the problem at hand. Some of the steps might involve additional experimentation, but overall benefits will be better understanding of the extraction process and better performance of the method. It should be emphasized that the optimization process has been evolving since the inception of the technique and may change in the future. Most SPME methods developed to date are used in combination with gas chromatographic separation and an appropriate detection method. Hyphenation with HPLC and other techniques discussed in Chapter 2 should be also considered. The discussion below is focused on the optimization of aqueous phase extractions, the most explored application of SPME at this time. The stages specific to other matrices are emphasized in the first three parts of Chapter 5.

4.1 Coating Selection

Chemical Nature of Analyte Determines Coating Type. As of fall 1996, Supelco has provided users with several coatings which included: three poly(dimethylsiloxane) (PDMS) films of different thickness (7, 30 and 100 μm), 85 μm poly(acrylate) (PA) and three 65 μm mixed phases, poly(dimethylsiloxane)/ poly(divinylbenzene) (PDMS/DVB), poly(ethylene glycol)/ poly(divinylben-

Table 4.1 Stages in SPME Method Development

Extraction strategy:
Coating selection
Derivatization reagent selection
Extraction mode selection
Agitation method selection
Selection of a manual vs automated system
Hardware:
Selection of separation and/or detection technique
Optimization of desorption conditions
Initial optimization:
Optimization of sample volume
Determination of extraction time profile in pure matrix
Determination of extraction time
Calculation of the distribution constant
Optimization of extraction conditions (pH, salt, temperature)
Calibration and validation:
Determination of linear dynamic range of the method for a pure matrix at optimum extraction conditions
Selection of the calibration method
Optimization of extraction conditions for heterogeneous samples
Verification of equilibration time, sensitivity, and linear dynamic range for complex sample matrices
Method precision
Method detection limits
Validation
Automation

zene) (Carbowax/DVB), and poly(ethylene glycol)/template poly(divinylbenzene) resin (Carbowax/TR). In the mixed phases, the DVB porous microspheres are immobilized onto the fiber by using either Carbowax or PDMS as a glue to hold them together. In addition, the pores of the template DVB polymer are uniform resulting in less adsorption discrimination as a function of analytes molecular weights.[1] These coatings satisfy many applications needs directed to the analysis of organic compounds. Figure 4.1 illustrates the general guidelines which can be used to determine the choice of the coating for a given application. Typically, the chemical nature of a target analyte determines the type of coating used. A simple general selection rule, "similar attracts similar" applies here. Knowledge of other extraction and separation techniques is helpful as well. Since only general coatings are available to date, the selectivity which can be obtained, is based primarily on polarity and volatility (molecular weight) differences among molecules. Figure 4.1 emphasizes the selection of coating according to these parameters.

PDMS is the Most Useful Coating. The most popular coatings to date are PDMS phases, and whenever possible they should be used. Advantages of

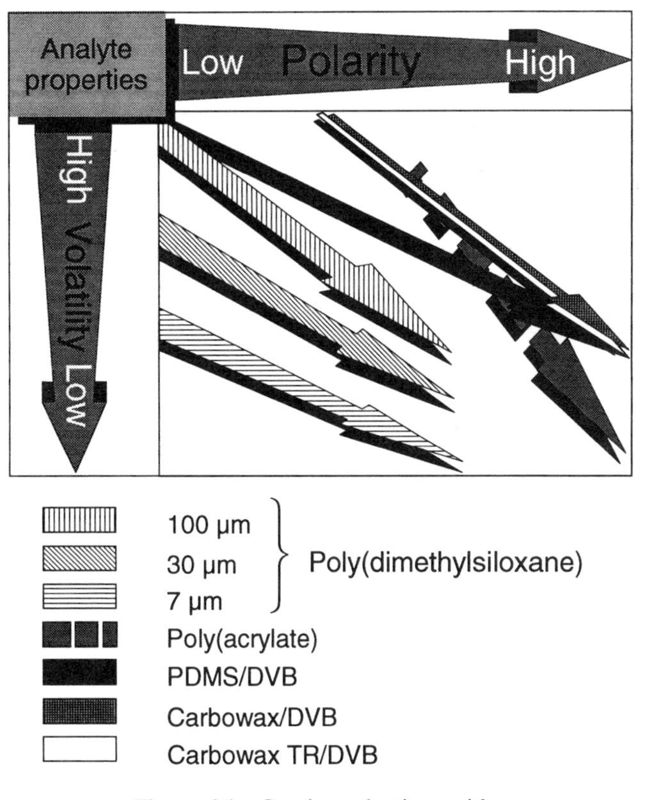

Figure 4.1 Coating selection guide.

these phases for SPME applications are similar to the advantages in their use as GC stationary phases. They are very rugged liquid coatings which are able to withstand high injector temperatures, up to about 300°C. PDMS is a nonpolar phase and it extracts nonpolar analytes very well. However, it also can be applied successfully to more polar compounds, particularly after optimizing extraction conditions. An additional advantage of using this phase, is associated with the fact that the appropriate distribution constants for organic compounds in air, can be conveniently estimated by considering retention times on the PDMS column,[2,3] since the same polymers are used as stationary phases. For aqueous sample extraction, the distribution constant also can be predicted by using the retention times combined with Henry's constants, characteristic for a given analyte (refer to Section 5.2)[4].

The Thinnest Coating which Achieves Required Detection Limits Should Be Used. Always consider that both the coating thickness and the distribution constant determine the extraction time; the same parameters also determine sensitivity of method. Therefore, some compromise needs to be reached at the beginning of method optimization. The decision about coating thickness is made after considering the distribution constant expected for given analytes and the sensitivity requirements placed on the method. As a general rule in direct aqueous extraction with magnetic stirring, a 100 μm PDMS coating provide equilibration times of less than one hour for compounds which have estimated distribution constant less than 10,000. For larger K, thinner PDMS coatings should be considered. For example, with PAHs and PCBs, a 7 μm PDMS film might be sufficient since the distribution constants are very large.[5] For polar compounds extracted using PDMS and polar phases, the distribution constant is typically less than 10,000 and therefore, there is no need to consider the use of thin phases.

PA Phase is Suitable for Polar Analytes. The PA phase, on the other hand, is suitable for more polar compounds, such as phenols.[6] It is a low density solid polymer at room temperature, which allows analytes to diffuse into the coating, but the diffusion coefficients are lower compared to PDMS, resulting in longer extraction times for volatile analytes in the headspace.

Mixed Phase Coatings are more Suitable for Volatile Compounds. Mixed phase coatings have complementary properties compared to PDMS and PA. Since the majority of interaction is determined by the adsorption process on porous poly(divinylbenzene) particles, they are more suitable for more volatile species, and the resulting distribution constants are typically higher compared to PDMS. Changing the "glue" from PDMS to Carbowax results in different selectivity toward polar compounds, such as ketones and alcohols. The adsorption times are typically shorter for gaseous samples compared to 100 μm PDMS, since the analytes do not need to diffuse through the liquid polymeric phase. The main disadvantages of solid sorbents compared to liquids are associated with smaller linear dynamic range and displacement effect.

DVB Template Resin is Designed to Reduce Molecular Weight Discrimination. Frequently, target analytes have polymeric structures that vary in chain length. If extracted by porous polymers, the amount extracted will vary as a function of their size in relation to the pore dimension. This is illustrated in Figure 4.2, where a group of alkylphenol ethoxylate nonionic surfactants is extracted using various DVB based polymeric coatings. The template resin exhibits uniform extraction efficiency which results in an extraction mixture representing the original distribution of oligomers in the sample (Figure 4.2a). Use of other DVB sorbents produces discrimination against high molecular weight substances, which indicate that some pores are too small to accommodate larger analytes (Figure 4.2b,c).

Method Development for Groups of Analytes Requires the Primary Consideration Be Given to the Most Difficult Analytes. Optimum coating type for a broad range of compound characteristics requires experimentation with different fibers. For example, as shown in Figure 4.3, if both organophosphorus and organochlorine pesticides are to be extracted from water samples, PA would be a phase of choice compared to PDMS since it can extract all the

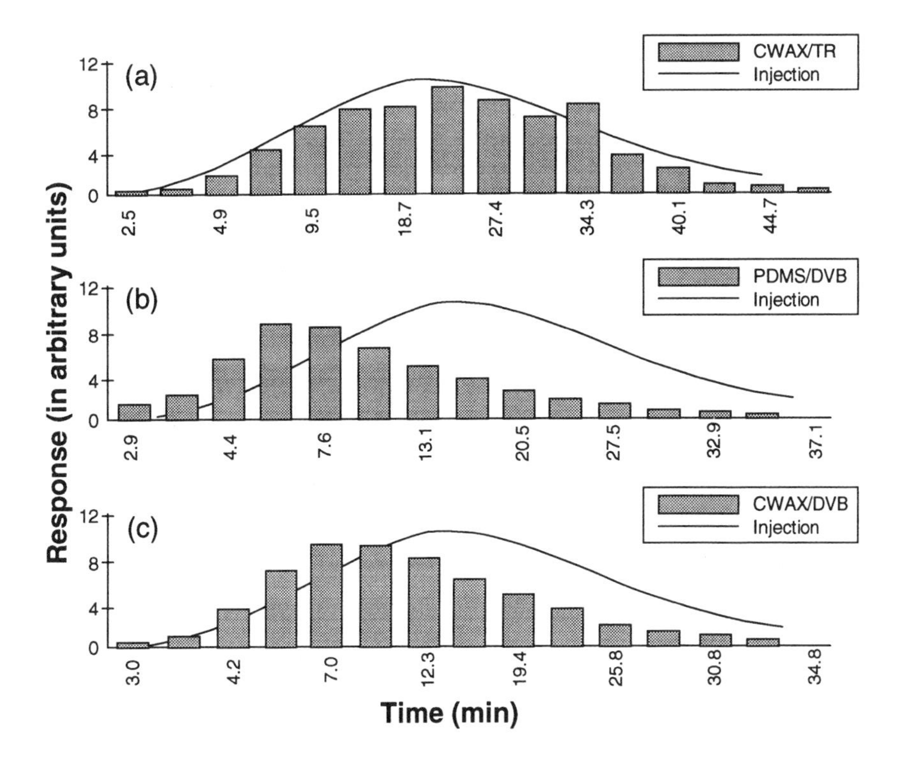

Figure 4.2 Comparison of ethoxamer distribution between direct injection of Triton X-100 standard and extraction of Triton X-100 from water using mixed phase fiber coatings.

analytes with acceptable sensitivity. PDMS is a better coating for many pesticides, but it fails to extract several others sufficiently well.[7]

Sometimes Coatings Must Be Custom-Made. Commercially available coatings might not be able to address certain needs. This gives researchers an opportunity to design custom fibers by using sorbing materials available in the laboratory or, when necessary, by working closely with polymer or organic chemists. This situation might occur frequently when coating with specific properties need to be applied, such as bioaffinity or molecular recognition coatings. It should also be remembered that some coatings are available in other forms (e.g., hollow fiber membranes). Figure 4.4 shows the separation of polar compounds extracted from fuel using a commercially available Nafion.[8] Techniques used to attach the films to fibers were discussed in Chapter 2.

4.2 Derivatization Reagent Selection

Derivatization performed before and/or during extraction can enhance sensitivity and selectivity of both extraction and detection, while postextraction

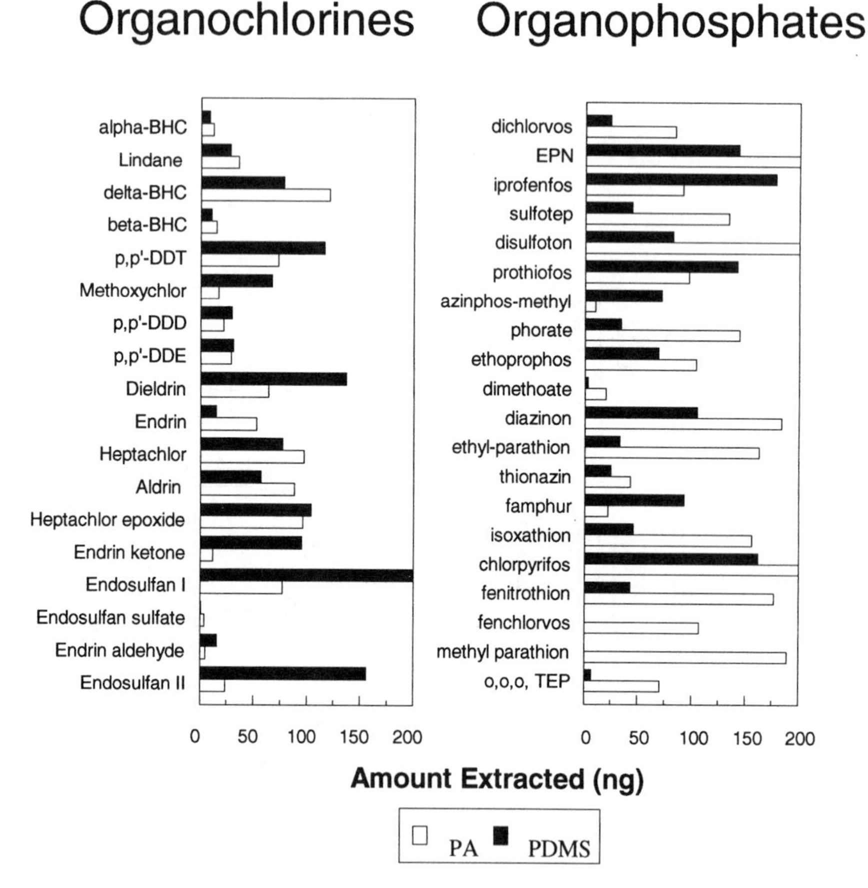

Figure 4.3 Performance of PDMS and PA coatings for extraction of organochlorine and organophosphorus pesticides from pure aqueous matrix spiked at 50 μg/L each.

methods can improve only chromatographic behavior and detection. The incorporation of the additional step of derivatization typically complicates the SPME method, and should only be considered when necessary. However, in cases where highly selective and/or sensitive analysis is required, this approach may be the appropriate one. A derivatization reagent can be used to enhance both selectivity and sensitivity of determinations. Selective reactions producing specific analogues will result in less interference during quantitation. This approach can be used for analyte determination in very complex matrices. Also, specific analogues can lead to enhanced detection if the introduced moieties have an unique property able to enhance detection. For example, it has been known for some time that 1-(pentafluorophenyl)diazoethane converts carboxylic acids directly in an aqueous matrix to pentafluorophenyl ester derivatives, which yield high sensitivity ECD detection and result in more

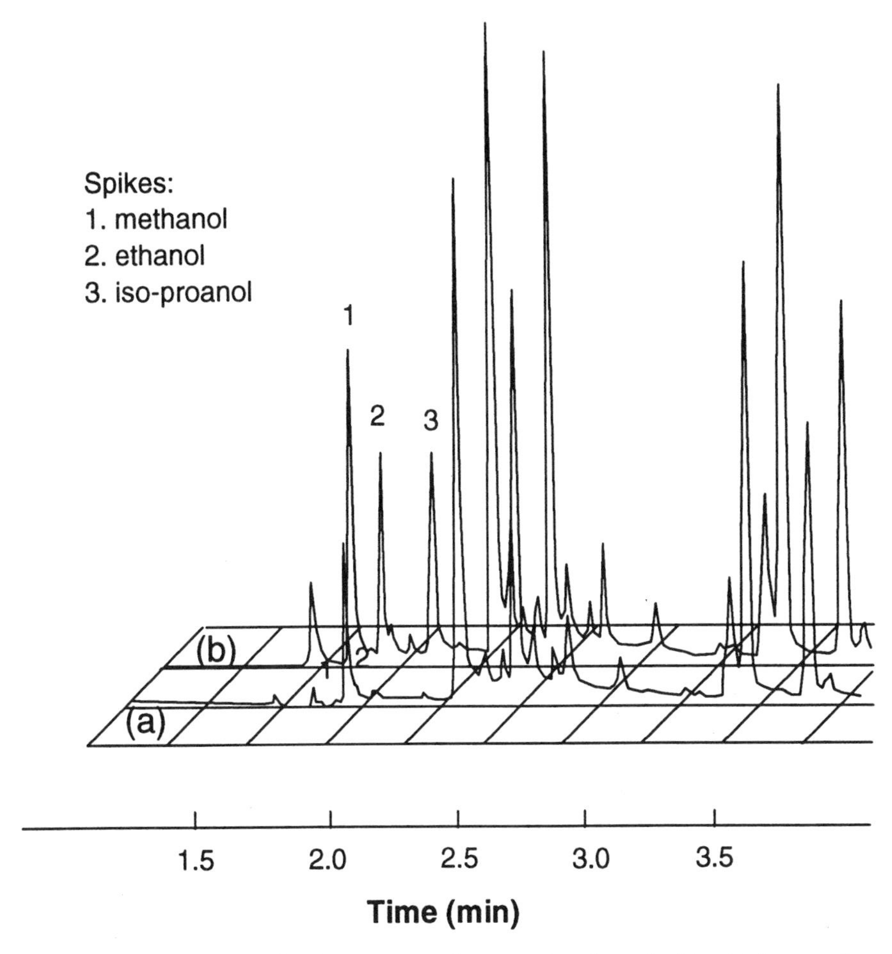

Figure 4.4 Separation of polar compounds extraction from fuel by Nafion coating: (a) chromatogram of unleaded gasoline, (b) chromatogram of unleaded gasoline spiked with 333 ppm each of methanol, ethanol, and isopropanol.

selective analysis.[9] This process can be adopted for SPME applications.[10] Various implementations for the derivatization processes to enhance SPME performance are discussed in Chapter 2. SPME/derivatization has been reported in the scientific literature, but no specific SPME commercial product yet supports this new, interesting extension of the technology. In many cases, derivatization reagents designed specifically for liquid extraction procedures can be used in SPME. For example, sodium tetraethylborate has been used to facilitate analysis of metals such as mercury and lead as well as to speciate organometallic compounds by SPME/GC.[11,12] In some cases, it might benefit researchers to consider designing specific reagents for unique target analytes.

The Most Powerful Implementation of the Derivatization Procedure Involves the Use of Doped Fibers. Figure 4.5 illustrates a chromatogram resulting from exposure of the PA coating containing 1-pyrenydiazomethane to the headspace of a sewage sample.[13] All volatile carboxylic acids are detected. Direct SPME, without derivatization, is unable to extract them in sufficient amounts because no selective films for these analytes are designed as of yet. The interesting features of such an approach is its high sensitivity and its compatibility with field analysis, since the derivatization reagent as well as derivatized analogs are not volatile. The doped fibers not only are able to facilitate spot sampling, but can also measure long-term exposure of a particular object or organism to a given contaminant, since analytes are continuously accumulated onto the fiber until the experiment is completed.

4.3 Extraction Mode Selection

Selection of the extraction mode is made by considering the sample matrix, analyte volatility and its affinity to the matrix (see Table 4.2). For very dirty samples, which contain substances able to damage the fiber coating, or change its properties through adsorption, the headspace or fiber protection mode should be selected. For clean matrices, direct sampling can be considered. However, for more volatile analytes, the headspace extraction method is still preferred due to faster equilibration times. Membrane protection should be used only for very dirty samples in cases where neither of the first two modes can be applied. Headspace SPME can generally be applied for analytes of medium to high volatility. For aqueous samples, it is frequently difficult to use this method well with very polar analytes, such as acids and bases, which have a high affinity toward the matrix. This situation can be improved by changing the pH of the matrix as discussed below. Application of the headspace SPME methods can be extended to semivolatiles and analytes strongly bound to the matrix by increasing the extraction temperature, also discussed below.

Equilibrium Versus Exhaustive Extraction. An important decision during method development is to choose between an equilibrium or an exhaustive

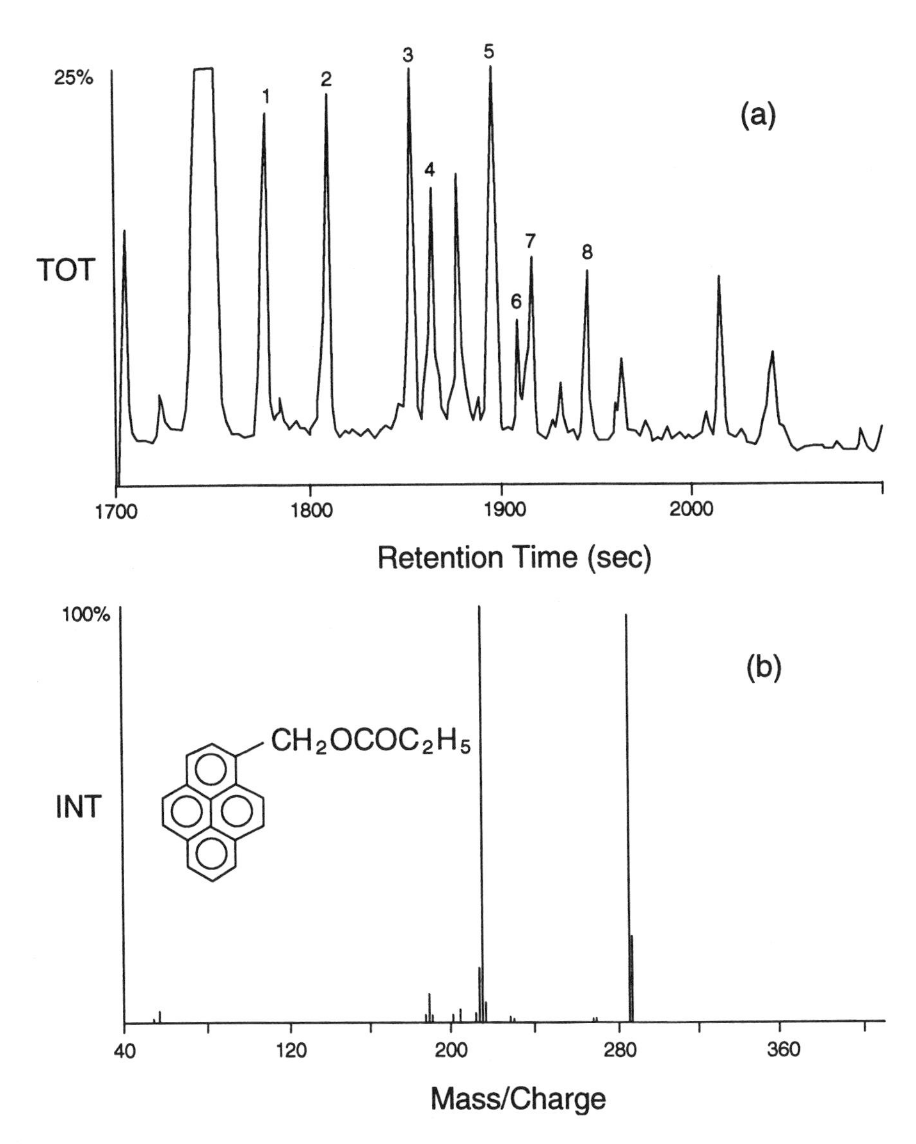

Figure 4.5 (a) Reconstructed GC/MS chromatogram indicating short chain fatty acids in a real sewage sample obtained with a PA-coated fiber. The sample was found to contain the following: 1, acetic; 2, propionic; 3, isobutyric; 4, butyric; 5, pivalic; 6, isovaleric; 7, valeric; and 8, hexanoic acids. The peaks correspond to pyrenylmethyl esters of these acids. (b) An example of the mass fragmentogram and the structure of propionic acid/PDAM ester.

Source: Adopted with permission from ref. 13.

Table 4.2 Sampling Mode Selection Criteria

Sampling Mode	Analyte Properties	Matrices
direct	medium to low volatility	gaseous samples, liquid (preferably simple)
headspace	high to medium volatility	liquid (including complex), solids
membrane protection	low volatility	complex samples

extraction approach. Any sample preparation technique can be adapted to perform equilibrium or exhaustive extraction. In SPME, it can be accomplished in several ways, the most universal approach is to cool the fiber coating.[14] Equilibrium methods have many advantages over exhaustive methods. An equilibrium method is always much simpler and less expensive. For example, in the cooled fiber approach, the fiber temperature needs to be kept low and constant during the experiment. Equilibrium methods are more selective because they take full advantage of differences in extracting phase/matrix distribution constants to separate target analytes from interferences. In exhaustive extraction approaches, selectivity is sacrificed to obtain quantitative transfer of target analytes into the extracting phase. Therefore, equilibrium methods typically do not require a cleanup step, while exhaustive methods frequently do.

Equilibrium approaches can be conveniently applied directly to field analysis. The need for a separate sampling step is eliminated when the target analytes are allowed to equilibrate with an extracting phase placed directly in or above a targeted area or object. This application of equilibrium methods not only shortens overall analysis time but also prevents loss of analytes and contamination from the sampling container. Despite the clear advantages of the equilibrium methods, however, the exhaustive approaches are frequently chosen even for screening investigations. Although many investigators believe that exhaustive extraction will reduce experimental errors, "exhaustive extraction" really describes the recovery of spikes, not native analytes. Thus, if strongly bound native analytes are not released from the matrix, they are not measured. If the extraction conditions are such that the native analytes are released quantitatively from the heterogeneous matrix, or if the matrix is homogeneous, equilibrium and exhaustive methods can give accurate results. One advantage of the exhaustive extraction approach is that, in principle, it does not require internal surrogate standards since all the analytes of interest are transferred to the extracting phase. On the other hand, the equilibrium approach usually needs surrogates or standard addition to compensate for matrix-to-matrix variations and their effects on distribution constants. In practice, however, the standards are also used to verify quantitative partitioning.

4.4 Agitation Technique Selection

The effectiveness of the agitation technique determines the equilibration times of aqueous samples. Equilibration times for the analysis of gaseous samples are fast and frequently limited only by the diffusion of analytes in the coating, particularly if flowing samples are used. A similar situation occurs when very volatile analytes are analyzed by the headspace technique. In this situation, most of the analytes are in the headspace resulting in relatively fast extraction times even when no agitation is used. However, in most cases, agitation is required to facilitate mass transport between the bulk of the aqueous sample and the fiber. Table 4.3 summarizes the properties of several agitation methods which have been tested with SPME.

Care must be taken when using a magnetic stirrer to ensure that the rotational speed of the bar is constant and the base plate is thermally isolated from the vial containing the sample. Magnetic stirring is most commonly used in SPME experiments since it is available in the majority of analytical laboratories and can be conveniently used with all three SPME sampling modes. Extraction is efficient when fast rotational speeds are applied. Figure 4.6 shows the effect of stirrer rotational speed on the absorbtion of the equilibration time profile for toluene.[15] The equilibration time progressively decreases with rpm increase, and for 3,330 rpm, it is about 200 seconds. Some caution is advised when purchasing a magnetic stirrer for use with SPME. Frequently, the rotational speed of the magnetic bar is not controllable, or constant, which might cause variation in agitation conditions during the extraction and change the equilibration times. The net effect could be poor measurement precision. Also, the base plate may heat up during stirrer operation, resulting in changes of the distribution constant, which again affect reproducibility of the measurement. Intrusive stirring can improve the agitation further,

Table 4.3 Agitation Methods in SPME

Method	Advantages	Disadvantages
static (no agitation)	simple, performs well for gaseous phase	limited to volatile analytes and headspace SPME
magnetic stirring	common equipment, good performance	requires stirring bar in the vial.
intrusive stirring	very good performance	difficult to seal the sample
vortex/moving vial	good performance, no need for a stirring bar in the vial	stress on needle and fiber
fiber movement	good performance, no need for a stirring bar in the vial	stress on needle and fiber, limited to small volume
flow through	good agitation at rapid flows	potential for cross-contamination, requires constant flows
sonication	very short extraction times	noisy, heats the sample

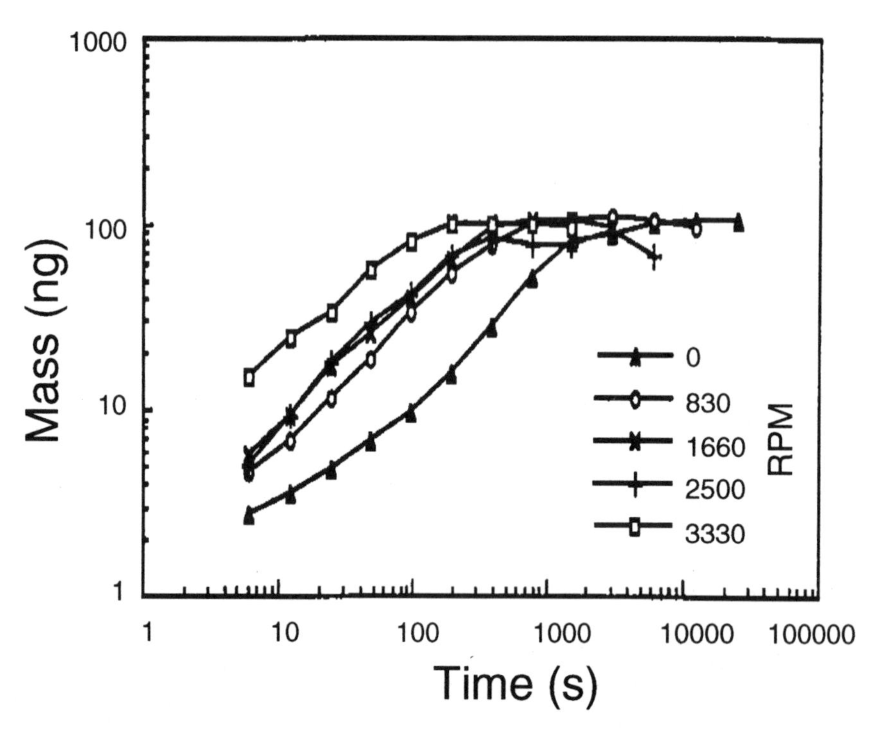

Figure 4.6 Extraction time profile corresponding to direct extraction with magnetic stirring of 100 ppb solution toluene using 56 μm PDMS coating.

Source: Adopted with permission from ref. 15.

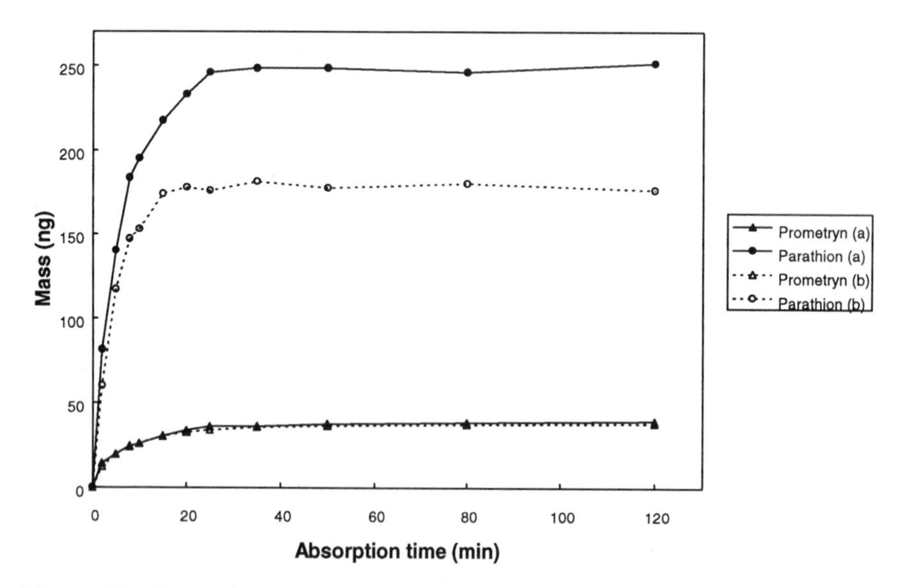

Figure 4.7 Comparison of the absorption time profiles obtained for parathion and prometryn (each 300 ppb in water) using (a) magnetic stirring with 40 mL vials and (b) the needle vibration technique in 2 mL vials.

but it requires a direct connection between the stirrer and the motor, which is difficult to seal.

Alternatives to magnetic stirring should be considered, particularly for automated operations. The needle vibration technique uses an external motor and a cam to generate a shaking motion of the fiber and the vial. This technique has been implemented by Varian in the design of a new agitated SPME autosampler. In the vortex technique, on the other hand, the vial is moved rapidly in a circular motion. Both techniques can provide good agitation, resulting in equilibration times similar to those obtained for magnetic stirring. However, for the needle vibration technique, good performance is generally limited to small vials and direct extraction mode. Both techniques can be conveniently automated to process a large number of samples since they do not require the use of foreign objects in the sample vials. Figure 4.7 illustrates equilibration time profiles obtained for two pesticides by needle vibration with 2 mL vials (Figure 4.7b) and for magnetic stirring with 10 mL vials (Figure 4.7a). The equilibration times are comparable for prometryn. The equilibration time for parathion, characterised by larger K_{fs} is shorter when using the vibration method compared to stirring. However, the amount extracted at equilibrium is lower for the 2 mL vials due to smaller sample volume which results

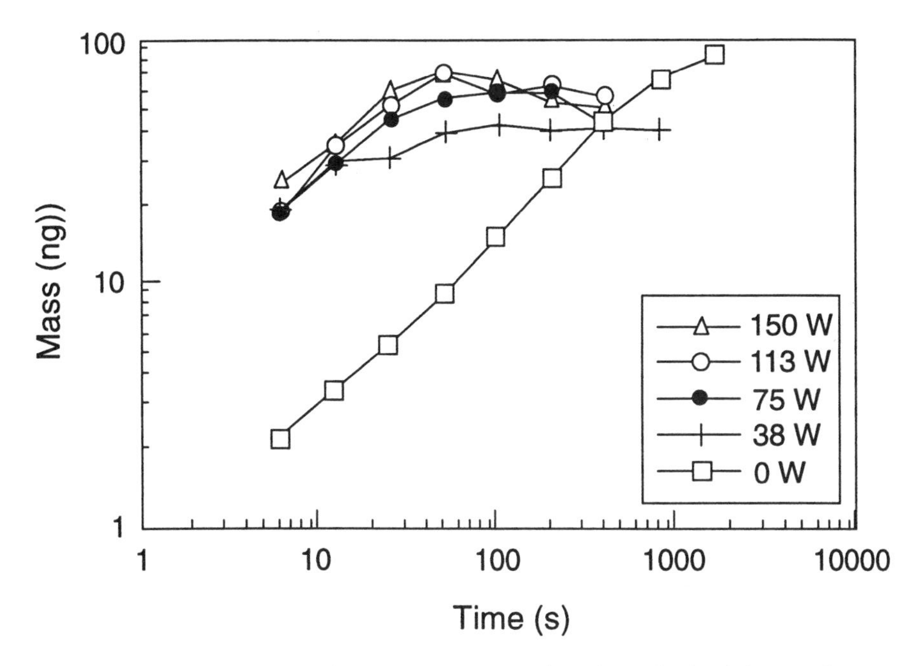

Figure 4.8 Effect of sonication power on extraction time obtained for continuous flow at 16 mL/min of aqueous sample containing toluene at 100 ppb. Coating thickness, 56 μm.

Source: Adopted with permission from ref. 15.

in faster equilibration (see Chapter 3). Flow-through techniques are very useful in continuous monitoring applications and also can be automated.[16] However, some additional flow metering devices may be required to ensure reproducible agitation.

The most efficient agitation method evaluated to date for SPME applications is the direct probe sonication, which can provide very short extraction times, frequently approaching the theoretical limits calculated for perfectly agitated samples. Figure 4.8 illustrates the dependence of the extraction time profile obtained for toluene as a function of the sonication power. The equilibration times are very short, approach the theoretical limit of 20 seconds, compared to 200 seconds for magnetic stirring (compare Figures 4.8 and 4.6). This technique has substantial drawbacks associated with the large amount of energy introduced into the system, which heats up the sample and, in some cases, can destroy the analyte.[12] The flow-through operation, however, eliminates most of the disadvantages.

4.5 Selection of a Manual vs Automated System

A manual system is a good starting point. A manual system generally provides more flexibility in the optimization of extraction parameters since common laboratory equipment can be adopted for use in SPME experiments, while automation typically requires dedicated equipment. An automated system, on the other hand, facilitates a faster optimization process. Also, experience with an automated system gained during optimization process would help the user to develop high volume applications. To date, the SPME autosampler has capabilities for both static operation and agitation but does not provide heating or temperature control. As SPME technology develops, the differences in capabilities between manual and automated setups are expected to diminish.

4.6 Selection of Separation and/or Detection Technique

Sample complexity and the selectivity of the extraction process define the requirements for separation and quantitation instrumentation. In most applications, the general coatings, such as PDMS have been used and therefore very limited selectivity has been obtained during the extraction process. Therefore, the demands on the separation/quantitation technique are very high. Most SPME applications have been developed for gas chromatography, but more recently, commercial interfaces to HPLC have been designed (see Chapter 2). Future SPME capabilities are expected to include coupling to capillary electrophoresis (CE) and supercritical fluid chromatography (SFC). The complexity of the extraction mixture determines the proper quantitation device needed. Frequently, mass spectrometers as detectors are required for complex environmental and biological samples. As selective coatings become available, the direct coupling to MS/MS[17] and ICP/MS becomes practical.

4.7 Optimization of Desorption Conditions

A narrow bore GC injector insert is required (see Table 4.4). Standard chromatographic injectors, such as the popular split/splitless types, require large-volume inserts to accommodate the expansion of the evaporated solvent introduced during injection. Large i.d. (3–5 mm) glass or fused silica tubes used for this purpose produce very low linear flow rates in the injector, resulting in a slow transfer of volatized analytes onto the front of the GC column (Figure 4.9a). The split opening allows injection band sharpening and removal of remaining solvent vapors. In SPME introduction, no solvent is present and therefore the split is unnecessary. In fact, for optimum sensitivity, the split needs to be closed during desorption to transfer all analytes onto the front of the GC column. It should be emphasized that the split must also be closed when calibrating the detector response to ensure that all analytes reach the detector. To facilitate sharp SPME injection bands, the analytes desorbed should be removed rapidly from the injector. This can be accomplished by generating high linear flow rates of the carrier gas around the coating. Since volumetric flow rates are fixed at optimum operation of gas chromatographic separation (about 1 mL/min), the practical way to achieve high linear flows is to reduce the diameter of the injector insert, matching it as closely as possible to the outside diameter of the coated fiber (Figure 4.9b). Narrow bore inserts of 0.8 mm i.d. are available from Supelco for a range of GC instruments.[18] Calculations show that it takes close to a minute to flush one void volume of carrier gas from the 3 mm i.d. insert (Figure 4.9a), while it lasts only a few seconds for 0.8 mm inserts (Figure 4.9b). Further reduction of the insert i.d. will facilitate even faster transfer.[19]

Sharpening of the Injection Band Can Be Accomplished by Using a Thick Film Phase Column, Cryofocusing, or retention gap. The few seconds needed to transfer analytes onto the front of the column might be still too long to produce narrow bands for very volatile analytes. A sharpening, however, can be accomplished practically by using either a thick film stationary phase

Table 4.4 GC Injectors and Their Compatibilities with SPME

Split/Splitless	PTV	SPI
can be used for SPME in spitless mode, low volume insert required	can be used for SPME, low volume insert required	low internal volume, best suitable for SPME

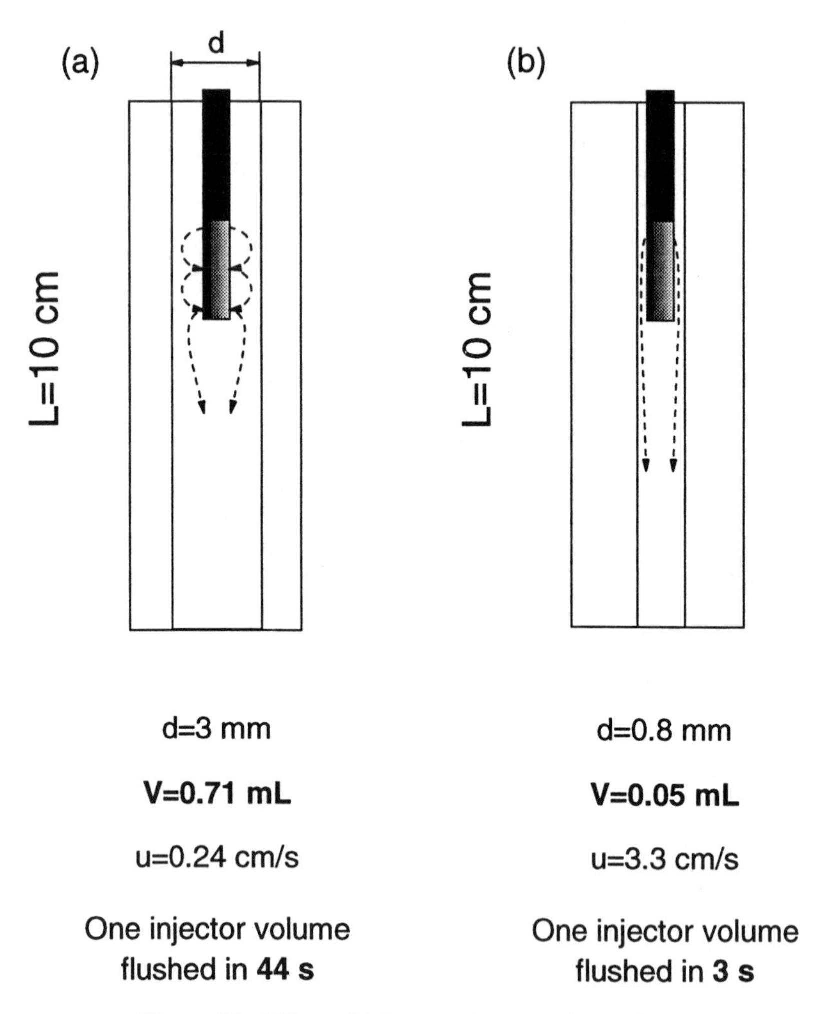

Figure 4.9 Effect of injector volume on desorption.

separation column or cryofocusing of the oven. It has been determined by a number of investigators that use of cryofocussing is generally not necessary for volatile compounds analysis and a 1 μm thick phase is sufficient to sharpen bands after desorption. It is always advisable to consider the application of a short length of deactivated narrow bore fused silica tubing as a retention gap. This will eliminate the broadening effect created by the negative temperature gradient existing in the column close to the injector. Figure 4.10 illustrates the operation principle of a retention gap.

Needle Exposure Depth Should Be Adjusted to Place the Fiber in the Center of the Hot Injector Zone. Figure 4.11 shows the typical temperature distribu-

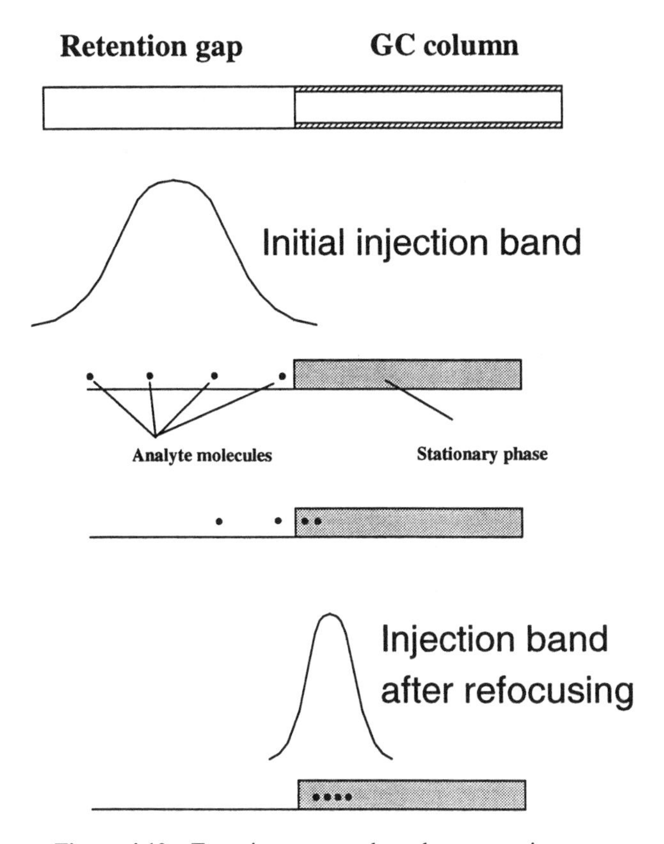

Figure 4.10 Focusing process based on retention gap.

tion in an injector. To ensure efficient, reproducible heating, the coated fiber needs to be placed in the center of the hot zone, which is typically located at the center of the insert.

The Fiber Needs to be Exposed Immediately after the Needle is Introduced into the Insert. Figure 4.12 shows the peak splitting effect obtained when the fiber remains in the needle for some time after introduction to the hot injector. The needle is heated rapidly after introduction to the insert and the analytes begin to desorb from the fiber, exit from the needle, and produce the first part of the split band. When the plunger is depressed, the remainder of the extracted mixture is introduced onto the column.

Temperature of the Injector and the Linear Flow Rate Around the Fiber Determine the Desorption Time. Theoretical desorption times are very short

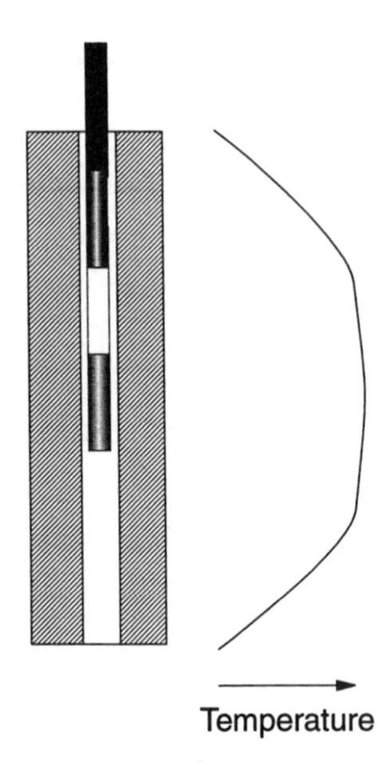

Figure 4.11 Typical injector temperature profile.

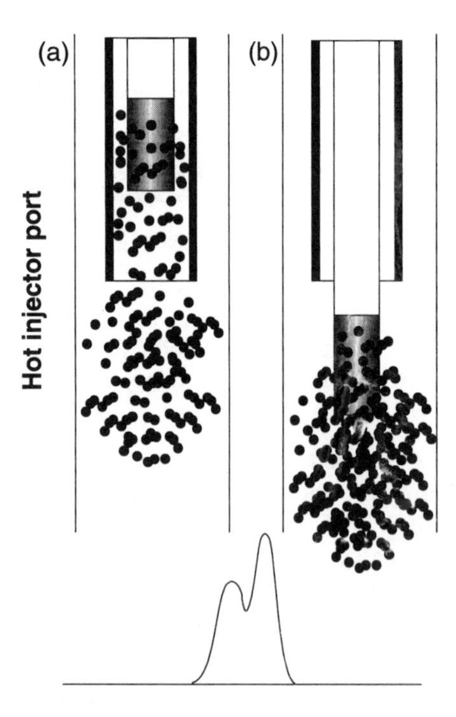

Figure 4.12 Formation of a split peak by slow fiber exposure to the injector.

Table 4.5 Effect of Temperature on Gas/Coating Distribution Constant, K_{fg}

	K_{fg}	
Compound	25°C	250°C
Benzene	300	0.4
o-Xylene	2,800	1.4
Undecane	24,000	13

since the diffusion coefficient of analytes in the coating increases and the gas/coating distribution constant rapidly decreases with temperature increase. Table 4.5 illustrates the dramatic decrease of the gas/PDMS distribution constant for three analytes when the temperature is increased from ambient temperature to 250°C, which is a typical injector temperature. The volatile analytes, benzene and xylene are fully removed from the coating at this high temperature. However, for lower boilers, such as undecane, the amount remaining at the coating at 250°C is still significant, which may lead to long desorption times if the flow of stripping gas around the fiber is slow. As discussed above, changing the inserts to narrow bore liners increases this flow dramatically. At high linear flow rates of the carrier gas, the injector temperature requirements are substantially reduced since the analytes do not need to be fully partitioned into the gas phase to be rapidly and exhaustively removed from the coating.

The Fastest Way to Optimize the Desorption Conditions is to Use the Maximum Allowable Coating Temperature as the Injector Temperature and then to Adjust the Desorption Time. A test can be performed using water or air standards containing target analytes. Following exposure to the sample, the fiber is placed inside the injector. After a predetermined time of desorption, the fiber is removed and sealed in the needle with a piece of septum. When the chromatographic run is completed, the fiber is immediately injected again, to determine carryover. Adjustment of the desorption time facilitates quantitative desorption in a single injection. Sometimes exceeding the desorption condition specified by the manufacture can help desorption of semivolatile analytes. However, application of too high temperatures can shorten coating lifetime and can result in the bleeding of the polymer, causing difficulty in separation and quantitation.

Frequently, Carryover and Poor Chromatography is Related to Pieces of Septum in the Injector Liner. The needle tip of SPME devices has a flat cut to allow convenient movement of the fiber in and out of the needle without damaging the coating. This, however, can lead to septum coring when the needle is introduced to the injector. This situation is aggravated by the use of narrow bore inserts, which direct the pieces of septum into the capillary column, resulting in plugging. Since the septum is typically made out of PDMS, presence of pieces of the polymer in the injector results in carryover and band broadening. To reduce this problem, the fiber inside the needle needs to be aligned with the end of the needle. Other solutions are to use prepunctured septa or septum-free injectors.

Parameters Which Control the Desorption Process in HPLC Interface are Analogous to GC Applications. In addition to temperature and flow, the appropriate selection of solvent to aid desorption is required. The simplest way to couple SPME with HPLC is to use the same desorption solvent as the mobile phase. In many instances, this is possible as in the case of PAH analysis.[20] In some cases, addition of the appropriate solvent to the interface will assist desorption. The linear flowrate should be maximized by choosing a small i.d. tubing as the desorption chamber. This is very important, since very low volumetric flow rates are used in HPLC, particularly with narrow bore columns.

Temperature also plays an important role in accelerating the desorption process as shown in Table 4.6. At ambient desorption temperature, significant amounts of surfactant remains in the coating, which is corrected when the temperature is raised to 135°C.[21] In addition, a judicious combination of the extraction and separation principles can lead to better desorption and selectivity. For example, extraction of proteins with an ion exchange fiber, followed by desorption using salt, and separation by reversed phase partition chromatography could be a powerful hyphenation of methods. It would

Table 4.6 Effect of Desorption Chamber Temperature on Carryover of Triton X-100 Using Carbowax/DVB Fiber

Number of Units in the Ethoxylate Chain, n	% Carryover	
	At 21°C	At 135°C
4	15	0
5	20	2
6	30	2
7	25	3
8	26	1
9	26	2
10	27	2
11	27	3
12	26	2
13	28	3
14	26	0

not only allow rapid desorption, but the combination of two separation principles (ion exchange and partitioning) would result in higher separation selectivity.

4.8 Optimization of Sample Volume

Volume of the Sample Should Be Selected Based on the Estimated Distribution Constant K_{fs}. The distribution constant can be estimated by using literature values obtained for the target analyte or a related compound with selected coating (see Chapter 5). If not available, an overnight direct extraction experiment should be performed to calculate the constant, using the formulas in Section 4.11. The sensitivity of the SPME method is proportional to the number of moles of analyte, n, extracted from the sample. For direct extraction method, it is given by:

$$n = \frac{K_{fs} V_f V_s C_o}{V_s + K_{fs} V_f}, \tag{4.1}$$

(see Chapter 3). As the sample volume, V_s, increases, so does the amount of analyte extracted, until the volume of the sample becomes significantly larger than the product of the distribution constant and volume of the coating (fiber capacity $K_{fs}V_f \ll V_s$). At this point, the sensitivity of the method does not increase with further increase in volume. In practice, the limiting volume can be calculated assuming the error of measurement, E:

$$V_s = \frac{100 K_{fs} V_f}{E} \tag{4.2}$$

For example, for 5% error, it can be estimated that $V_s = 20K_{fs}V_f$. This means that for K up to about 200 and 100 μm coating, the 2 mL vial is sufficient to give maximum sensitivity, while a 40 mL vial can be used for K smaller than 4,000, etc. Using sample volumes larger than the limiting value does not only maximize the sensitivity, but also results in better precision since the variation of the sample volume does not affect the results. Figure 4.13 illustrates this relationship graphically. The n_I corresponds to the amount extracted at infinite volume. Even for a moderate distribution constant of 1000, characteristic for many volatiles, and 100 μm PDMS coating (see Figure 4.13a), there is substantial variation in extracted amount when only a few mL sample is used. For example, increasing the volume from 1 mL to 10 mL results in sensitivity improvement by over 50%. The change is more dramatic for semivolatile analytes. The variation is less significant for thinner coatings (Figure 4.13b,c), but sample volume should still be considered when analytes characterized by large K, such as PAHs, are analysed. In practice, the sample size is determined by the dimension of the available vials. The discussion above should assist in the choice of the most appropriate vial size.

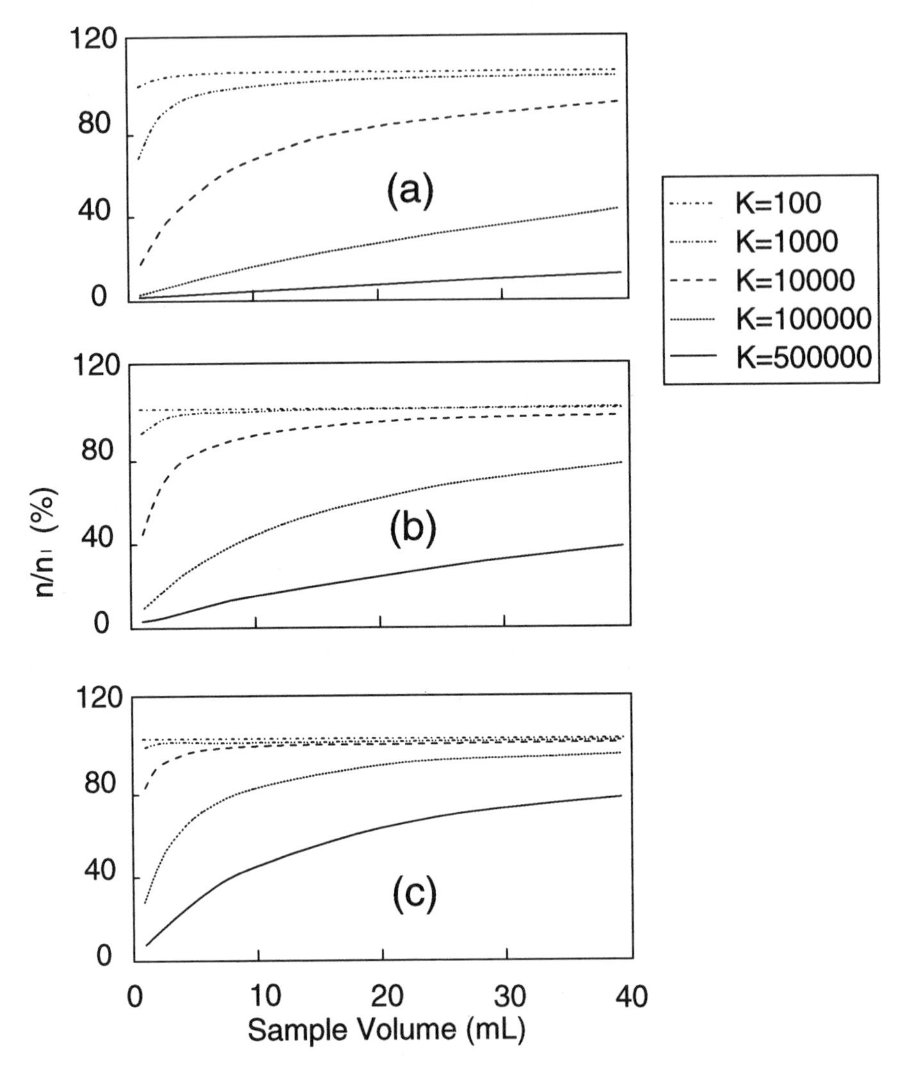

Figure 4.13 Dependence of n/n_l on V_s for (a) 100 μm fiber (V_f = 0.65 μL), (b) 30 μm fiber (V_f = 0.14 μL), and (c) 7 μm fiber (V_f = 0.028 μL) for direct SPME from small-volume samples in a two-phase system.

For High Sensitivity Headspace Extraction, the Volume of the Gaseous Phase Should Be Minimized. In headspace SPME, the situation is more complex compared to direct extraction as discussed earlier, since the analytes partition to the gaseous phase as well as to the coating. Very volatile compounds will prefer to accumulate in the headspace, resulting in a very substantial loss in sensitivity when the headspace is large. Figure 4.14 illustrates the loss of sensitivity for a volatile analyte when the headspace/sample volume ratio

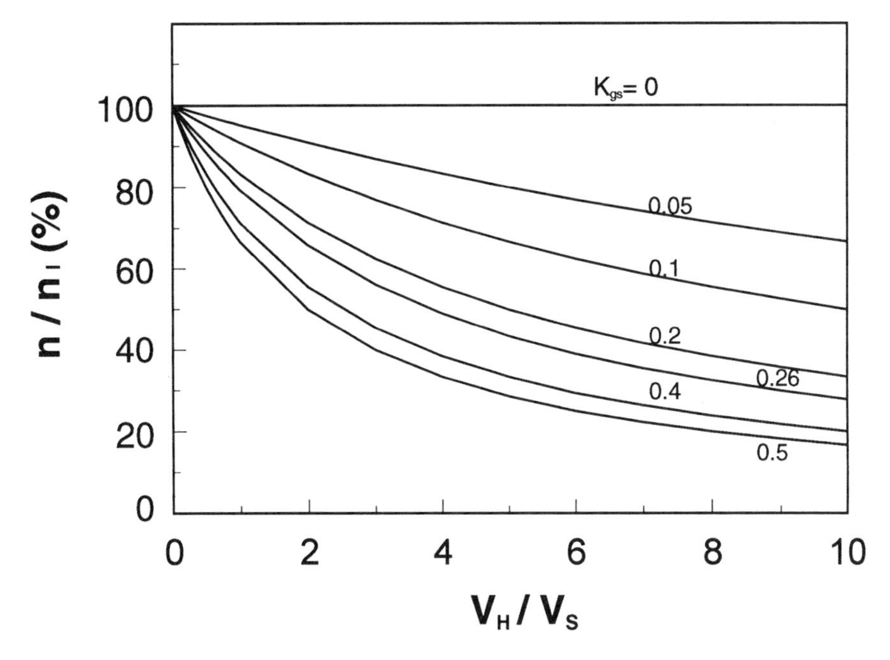

Figure 4.14 Dependence of relative extracted amount from headspace, n/n_I, on the headspace/sample volume ratio for analytes with different K_{gs} but identical K_{fs}, where n_I is the amount extracted at infinite volume and direct SPME.

increases. As a more practical example, Figure 4.15 considers extraction of three compounds: chloroform, 1,1,1-trichloroethane, and carbon tetrachloride from 4, 15, and 40 mL vials by a 100 μm fiber, typically used for the analysis of volatile compounds in sample headspace. n_I is the amount extracted at infinite volume and direct extraction. The following values of distribution constants have been used: $K_{gs} = 400$, 2500, and 1000, and $K_{hs} = 0.15$, 0.7, and 1.24, respectively.

It is interesting to note that n/n_I is always the largest for chloroform, which has the lowest K_{gs} value, even though its K_{fs} value is also the lowest. This is due to the fact that only a small fraction of the analyte is present in sample headspace; therefore equilibrium concentration of the analyte in the sample remains relatively high. The shape of the two other curves in Figure 4.15 is also very interesting. For low headspace volumes, n/n_I is higher for carbon tetrachloride than for 1,1,1-trichloroethane, which is consistent with the values of the distribution constants K_{fs}. However, as carbon tetrachloride has the largest K_{hs}, n/n_I drops faster for this compound than for 1,1,1-trichloroethane, as a result of which the two lines cross. This illustrates the significance of headspace volume on the analytical results. As expected, the biggest relative increase in n/n_I ($\sim$44%), when moving from small to large vials, is observed for 1,1,1-trichloroethane, with the highest K_{fs} value. For compounds with lower

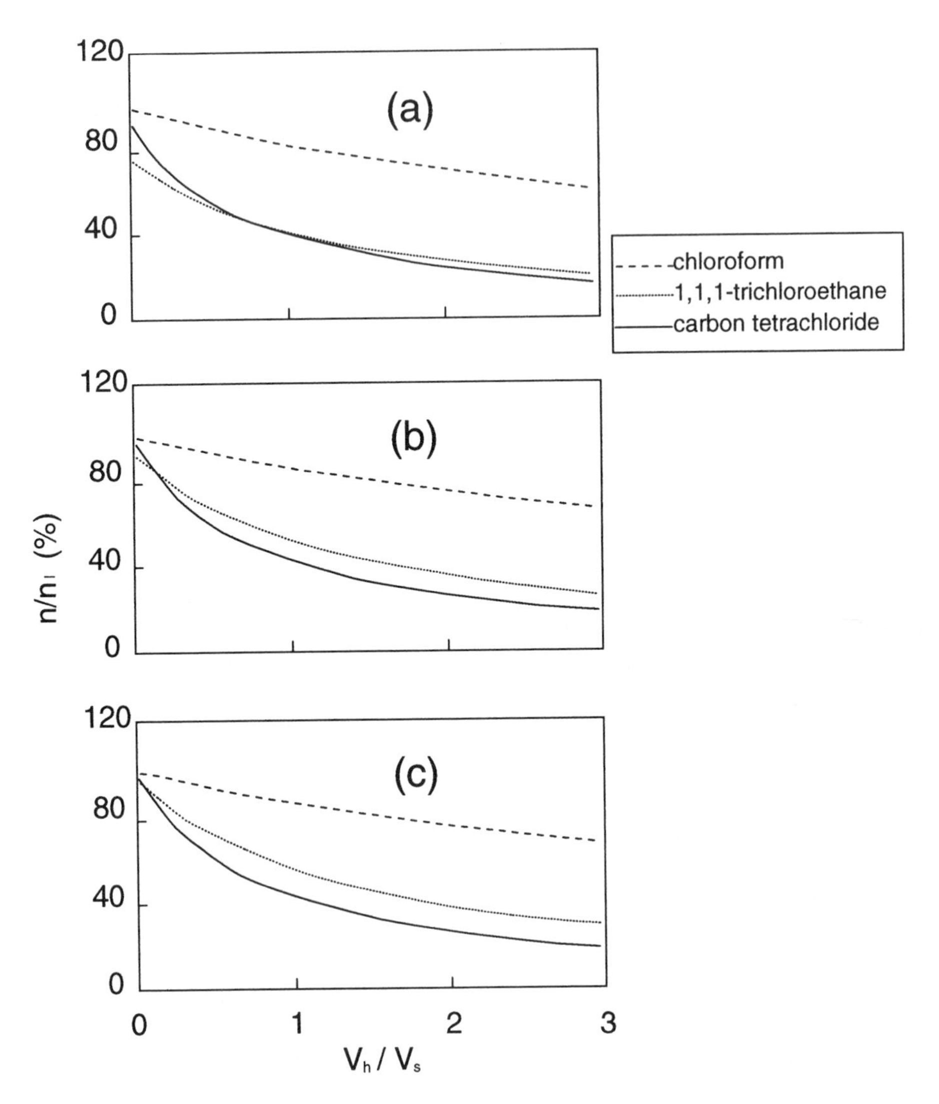

Figure 4.15 Effect of headspace volume on the amount of analyte extracted in a system of a constant volume of (a) 4, (b) 15, and (c) 40 mL by a 100 μm fiber for chloroform, 1,1,1-trichloroethane and carbon tetrachloride.

K_{fs}, the relative increase in the amount extracted is much lower (~16 and ~12% for chloroform and carbon tetrachloride, respectively). It is, therefore, the combination of K_{fs} and K_{gs} for a given compound that determines the magnitude of the sample volume effect on the amount extracted in three-phase systems involving headspace.[22]

For Rapid Extraction, the Headspace Capacity Needs to be Optimized. If the amount of analyte extracted onto the fiber is small relative to the amount of analyte contained in the headspace (headspace capacity), the amount of analyte required to be transferred from the aqueous or solid matrix to the headspace is of the same order of magnitude as the experimental error. Such conditions result in fast extraction, since the majority of analytes come to the fiber from the gaseous phase and there is little need for equilibrium to be re-established between the sample matrix and the headspace (see Chapter 3). These conditions can be represented by eq. 4.3:

$$\frac{K_{fs}V_f}{K_{hs}V_h} = \frac{E}{100} \tag{4.3}$$

where E is experimental error. For example, if we assume an experimental error of 5%, the capacity of the headspace ($K_{hs}V_h$) needs to be about 20 times larger than the capacity of the fiber ($K_{fs}V_s$). Under these conditions, extraction takes only a few minutes and is relatively independent of the agitation conditions, since the analytes are extracted from the small volume of the headspace (refer to Figure 4.20). The capacity of the headspace is controlled by both the volume of the gaseous phase and the headspace/sample distribution constant. $K_{hs} = K_{gs}$ can be optimized by adjusting the temperature and modifying the sample matrix, similarly to K_{fs}.

Another way to increase headspace capacity is to increase volume of the headspace (V_h). If the K_{hs} is known, the required volume can be calculated using eq. 4.3. However, in practice, the volume of the sample is limited to maximum vial size, which rarely exceeds 40 mL. It is emphasized, as discussed above, the increase in headspace capacity causes a significant loss in sensitivity. The discussion above can be used to decide about the vial size to be used in a given experiment. Also, it should be remembered that changing vial size during the optimization process, for example from 40 mL vials to 2 mL autosampler vials, will affect not only sensitivity but also equilibration times because of change in headspace volume.

The Vial Shape and Headspace Volume has an Effect on the Matrix-Fiber Mass Transfer Kinetics. Frequently, headspace conditions are such than the headspace capacity is low. In this case, the headspace volume and sample/gas contact area affects kinetics of headspace SPME, since the analytes need to be transported through the interface and the headspace, in order to reach the fiber. The smaller the gas phase is with respect to the sample, the more rapid is the transport of analytes from the sample matrix to the fiber coating (see Chapter 3). In static SPME, the vial cross section will determine the mass transfer rate between the sample and the headspace. Also, depending on the shape of the vial, more or less convection is produced for the same agitation technique. Long thin vials may be difficult to stir uniformly compared to larger diameter ones. For example, a sample contained in a standard 2 mL autosampler vial is difficult to agitate with magnetic stirring, because of the small diameter.

Sample Size Might Be Determined by Heterogeneity of the Sample. The discussion above indicates that even small sample volumes can frequently provide conditions close to optimum sensitivity. However, proper quantitation of heterogeneous samples (sludge, soil, etc.) may require a larger volume of material to correctly represent the investigated system. The required volume can be determined by using sampling statistics. Sample reduction techniques can be used to obtain manageable volumes.[23]

4.9 Determination of Extraction Time Profile in Pure Matrix

At this stage, most of the fundamental parameters have been determined for successful SPME measurements. The first experiments performed should be on a pure matrix such as dry air, pure water, or sand, to obtain a basic understanding about the kinetics of analyte transfer from matrix to fiber at set agitation conditions. The initial experiments will confirm the ability of analyte extraction by the chosen mode, and indicate if there is a need for further optimization. It will also give a good estimate of the detection limits which are to be expected for the analyte/matrix combination. The time extraction profile is obtained by preparing a set of vials containing samples and then extracting them for progressively longer periods of time. The extracted mass is determined by a selected method and an appropriate calibration procedure, which typically involve the injection of 1 μL of the target analyte mixture in a noninterfering solvent.

Caution must be taken to prevent moisture from being introduced into the needle. When performing experiments with aqueous samples, it is easy to introduce small amounts of water into the needle when agitation is used. In some cases, moisture can also condense inside the needle through capillary forces when high humidity solids and gases are analysed. Because the construction of the SPME interface is optimized for solvent-free operation, the presence of water interferes with its operation. Expansion of water vapor would produce loss of analytes in the septum purge line and, therefore, difficulties in proper quantitation. In the majority of cases, positioning of the needle far from the water surface is sufficient to overcome this problem. To eliminate it, the needle, after piercing the vial septum, can be withdrawn, allowing the septum to seal the fiber in the vial via the metal tubing, thereby isolating the needle from the sample (see Figure 4.16). To remove the fiber at the completion of extraction, the needle first pierces the septum to allow safe withdrawal of the fiber. To apply the same principle to direct extraction as shown in

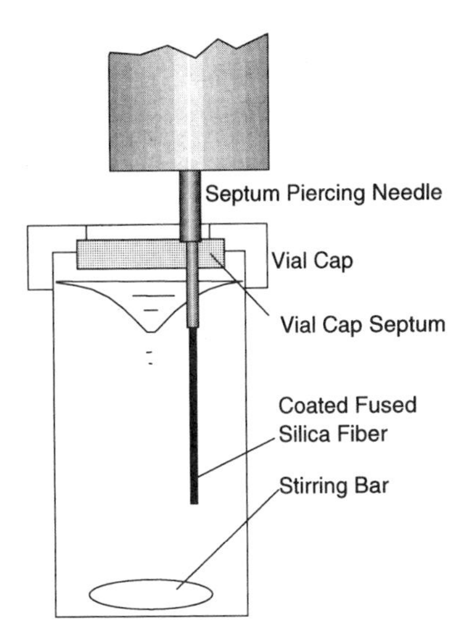

Figure 4.16 Experimental approach to protect against water in the needle.

Figure 4.16, a little headspace must always be present on the top of the vial. In addition Figure 4.16 illustrates the optimum position of the fiber to facilitate the fastest equilibration, as emphasized in Chapter 3. The highest linear velocity of a liquid sample is obtained at the position corresponding to half the distance between the center of the vial and the end of a stirring bar.

4.10 Determination of the Extraction Time

The objective of the SPME experiments, in principle, is to reach distribution equilibrium in the system. At this condition, the system is stationary, and therefore a variation of mass transfer does not affect the final results. The equilibration time is defined as the time after which the amount of extracted analyte remains constant and corresponds within experimental error to the amount extracted at infinite extraction time . Care needs to be taken when determining equilibration times since, in some cases, substantial reduction of the slope might be wrongly assigned as the equilibration point. This is particularly true in headspace SPME determinations when the rapid rise of the equilibration curve, corresponding to extraction from the gaseous phase, is followed by a very slow increase related to mass transfer of sample from water through the headspace to the fiber. Figure 4.17 illustrates an example of such a curve for naphthalene. The amount extracted increases rapidly in the first 5 minutes, but equilibration is not reached until much later. Determination

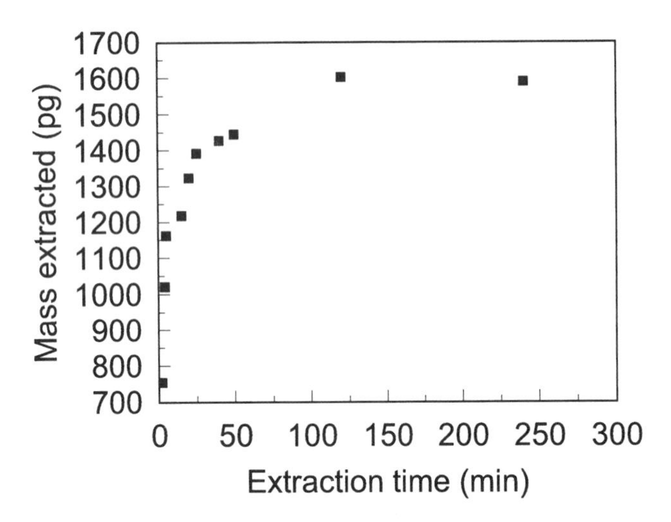

Figure 4.17 Extraction time profile corresponding to extraction of phenanthrene-d10 from water at 50°C under magnetic stirring. Fiber coating: 7 μm PDMS.

of the amount extracted at equilibrium allows calculation of the distribution constants. This information can frequently be used to compensate for any variation in extraction conditions which might occur during the experiment, for example, as in field monitoring.

When using a shorter extraction time compared to equilibration time, care must be taken to control the exposure time and the longest possible extraction time should be applied. When equilibration times are too long for a given application, shorter extraction times can be chosen for quantitation. Figure 4.18 shows the equilibration time profiles for *p,p'*-DDT (a) and dichlorovos (b). The equilibrium conditions for DDT are reached after about 2 hours, and the slope of the curve is small. Therefore a 1-minute deviation in the exposure of the fiber results in about the same absolute error independent of the extraction time. However, the relative error becomes smaller for longer times. For example, for 10 minutes of exposure, the relative error due to timing could be as high as 20%, but it drops to 5% for 50-minute extraction times since more sample accumulates on the fiber. When the equilibration curve rises rapidly, the timing is very critical for short exposure times. Figure 4.18b shows an example where a 1-minute error in 10 minutes sampling time results in a 50% relative error of determination. Autosampling devices can measure the time very precisely, and determination made using these devices can be done with very good precision, even when the equilibration conditions are not reached. However, constant convection and temperature in the system needs to be ensured to obtain reproducible data. This condition requires good temperature control and constant agitation and, for static conditions, a vibration-free environment.

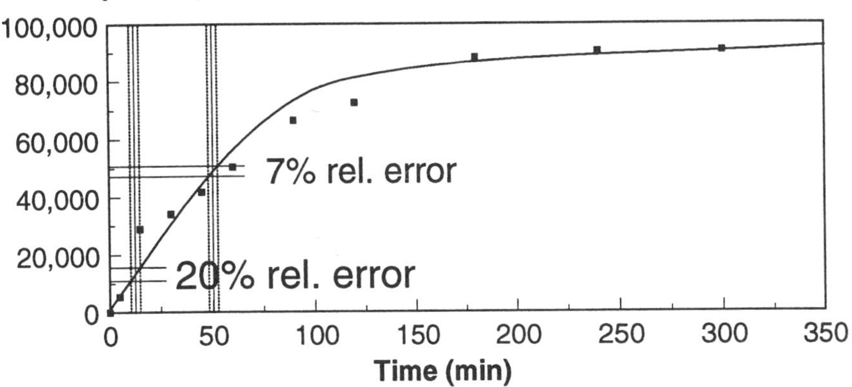

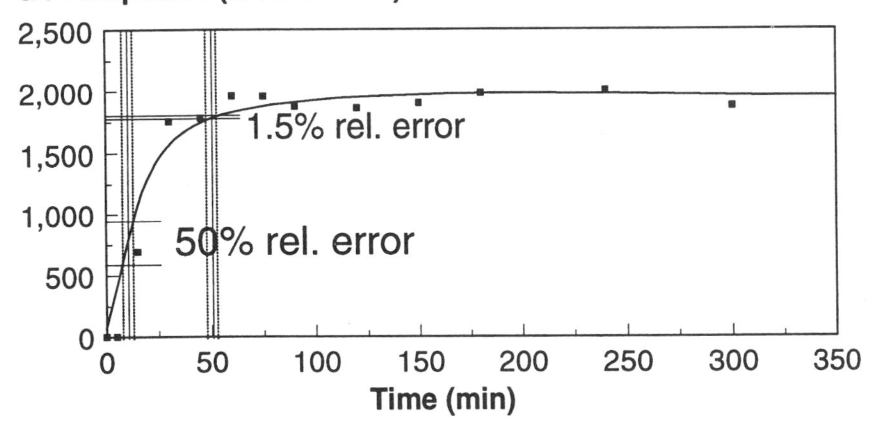

Figure 4.18 Selection of the extraction time based on extraction time profiles: (a) p,p′-DDT and (b) dichlorovos.

4.11 Calculation of the Distribution Constant

Target analyte's distribution constant for a pure matrix defines the sensitivity of the method. It is not necessary to calculate the fiber coating/sample matrix distribution constant, K_{fs}, if calibration is based on isotopically labeled standards or standard addition; the same is true even with external calibration when identical matrix and headspace volumes are used in a standard and in

Table 4.7 Ranges of K Values Obtained When the Amount of the Analyte Extracted by the Fiber n Determined Experimentally Falls Within ±5% (RSD) of the True Value, for Two Different Fiber Coating Thickness and Two Sample Volumes

100 μm			
2 mL sample		35 mL sample	
K	Range of K	K	Range of K
100	105–95	100	105–95
1,000	1,067–935	1,000	1,051–949
10,000	12,537–8,172	10,000	10,598–9,413
100,000	∞–36,190	100,000	115,748–86,928
1,000,000	∞–55,072	1,000,000	4,700,000–492,593
10,000,000	∞–58,104	10,000,000	∞–923,611
100,000,000	∞–58,426	100,000,000	∞–1,012,177
7 μm			
2 mL sample		35 mL sample	
K	Range of K	K	Range of K
100	105–95	100	105–95
1,000	1,051–949	1,000	1,050–950
10,000	10,574–9,434	10,000	10,504–9,496
100,000	112,903–88,785	100,000	105,422–94,622
1,000,000	3,500,000–558,824	1,000,000	1,093,750–913,462
10,000,000	∞–1,187,500	10,000,000	17,500,000–6,785,714
100,000,000	∞–1,338,028	100,000,000	∞–19,000,000

a sample. However, it is always advisable to determine K_{fs}, since this value gives more information about the experiment and aids optimization. K_{fs} can be used to predict the distribution of analytes in the system. This information allows the calculation of the headspace volume, sample volume, and coating thickness required to reach the desired sensitivity. Changes in K_{fs} caused by variation in extraction condition, such as temperature, sample volume, pH, and salt can be estimated theoretically by using the principles discussed in Chapter 3.

The distribution constant for the direct extraction mode can be calculated from the following equation, obtained from eq. 4.1:

$$K_{fs} = \frac{nV_s}{V_f(C_0V_s - n)} \tag{4.4}$$

where n is the amount of analyte partitioned into the coating. It is important to consider the sample volume when calculating the value of K_{fs}. Table 4.7 shows the ranges of values which are obtained using eq. 4.1 and assuming 5%

error of determination. To correctly estimate the distribution constants over 1000 with 0.1 mm thick PDMS fiber, volumes of the sample larger than 2 mL need to be used. For large K_{fs} (1,000,000), large hydrocarbons in water or air, the sample volumes need to be at least 35 mL. If the distribution constant is low and the total volume of the sample is high, a good estimate is given by:

$$K_{fs} = \frac{n}{C_0 V_f} \tag{4.5}$$

However, extreme caution is advised when using this equation since it assumes that, after extraction, the analyte concentration does not change significantly in the sample. If this equation is used by mistake for larger K_{fs}, the results would correspond not to K_{fs} but rather to the sample/coating phase volume ratio, V_s/V_f. This is because, at large K_{fs} and small sample volumes, most analytes are already extracted onto the fiber and a further increase of the value of K_{fs} does not change the amount extracted. Figure 4.19 illustrates this relationship graphically for different coating volumes and K_{fs}.

To obtain the coating/matrix distribution constant when headspace is present in the vial, knowledge of the matrix/headspace distribution constant is required in order to calculate K_{fs},

$$K_{fs} = \frac{n(K_{hs}V_h + V_s)}{V_f(C_0 V_s - n)} \tag{4.6}$$

K_{fs} can be obtained for aqueous samples from tables of Henry's constants, K_H:

$$K_{hs} = \frac{K_H}{RT} \tag{4.7}$$

For an analytes with unknown Henry's constant, the following equation can be used:

$$K_{fs} = \frac{nK_{fh}V_s}{K_{fh}V_f V_s C_0 - nK_{fh}V_f - nV_h} \tag{4.8}$$

Equation 4.8 is obtained by substituting $K_{hs} = K_{fs}/K_{fh}$ into eq. 4.6 and rearranging it. The fiber/gas distribution constant ($K_{fg} = K_{fh}$) can be found by extracting target compounds from air mixtures using a simple bulb experiment.[24] In addition, K_{fg} can be obtained directly from a chromatographic run when the stationary phase is made of the same material as the fiber coating. This has been discussed in Chapter 5. For compounds which have low vapor pressure, the amount of analyte present in the headspace is very small, especially if the volume of the headspace is kept to a minimum. Therefore, in this situation, headspace volume can be neglected altogether and eq. 4.1, which corresponds to direct extraction, can be applied.

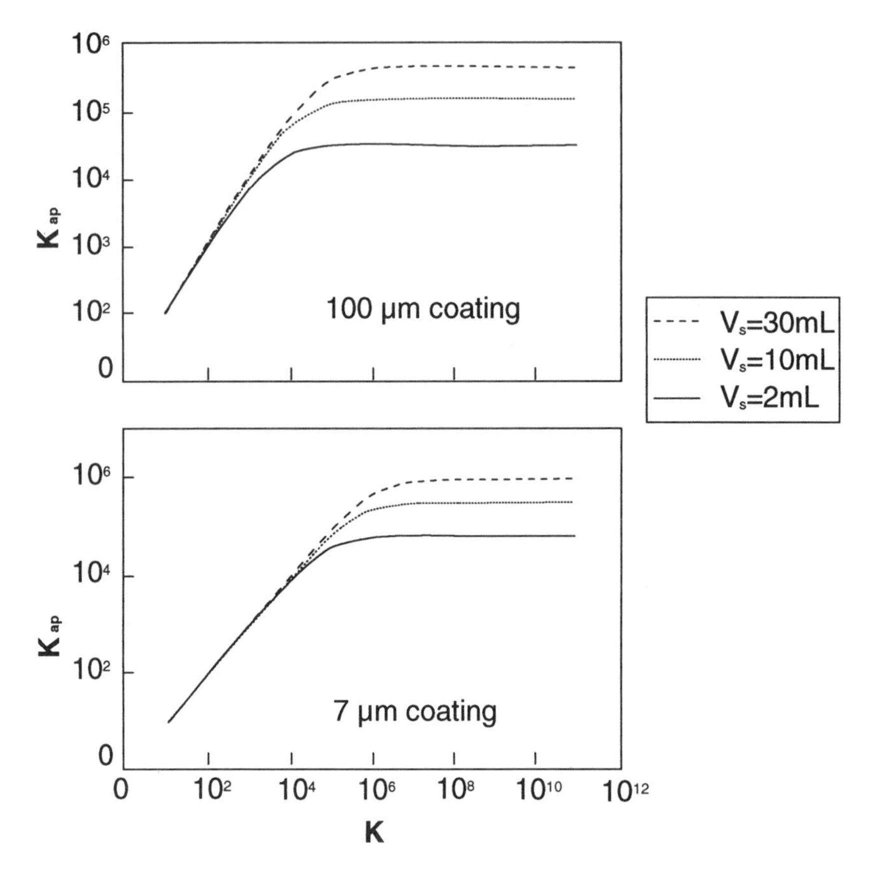

Figure 4.19 Relationship between $K_{ap} = \frac{C_f^\infty}{C_0}$ and $K = \frac{C_f^\infty}{C_s^\infty}$ for two different coatings (100 and 7 μm) and three sample volumes (2, 10 and 30 mL).

4.12 Optimization of Extraction Conditions

Temperature Affects Both Sensitivity and Extraction Kinetics. As emphasized in Chapter 3, an increase in extraction temperature causes an increase in extraction rate, but simultaneously a decrease in the distribution constant. In general, if the extraction rate is of major concern, the highest temperature which still provides satisfactory sensitivity, should be used. An internally cooled fiber SPME device eliminates the sensitivity loss, but this device is not commercially available at publication time. Figure 4.20a demonstrates the effect of temperature on extraction of methamphetamine from water. At room temperature, Figure 4.20a, equilibration takes several hours. When the temperature is raised to 60°C, equilibration time is less than 20 minutes for both compounds. When the temperature is further increased to 73°C, the

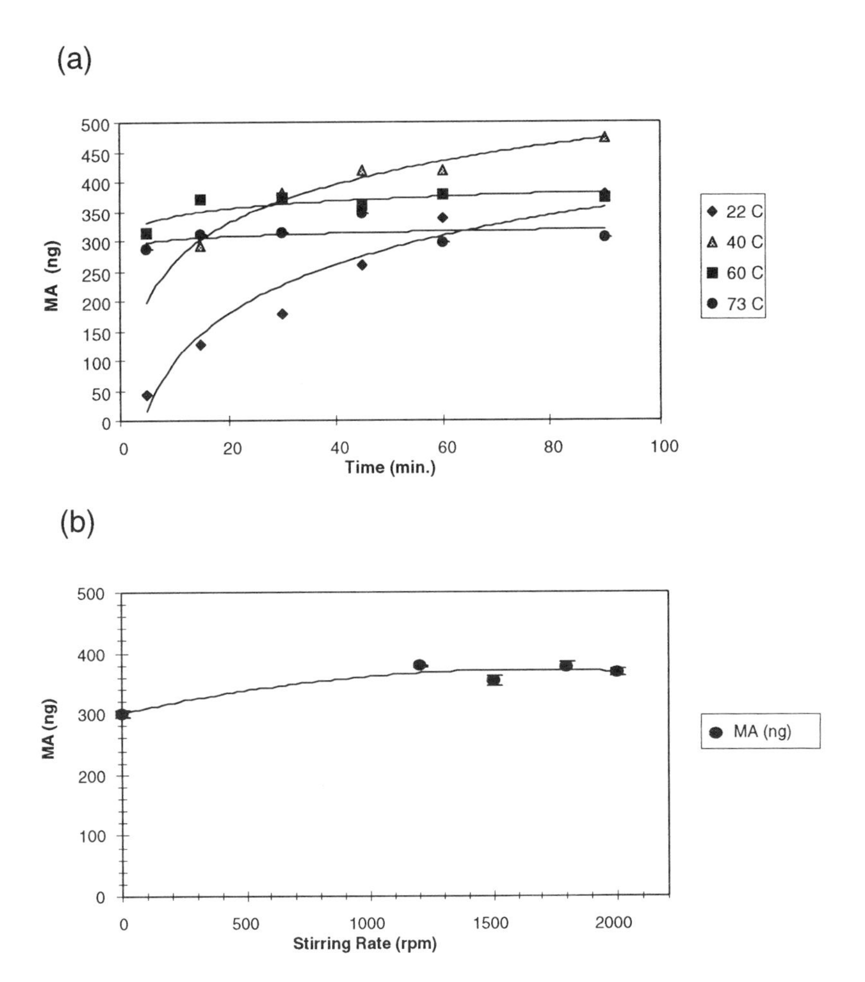

Figure 4.20 (a) Temperature dependence of the absorption time profile obtained for methamphetamine (22, 40, 60, and 73°C). (b) Amount of methamphetamine adsorbed as a function of the stirring rate at 60°C.

extraction times are only a few minutes (Figure 4.20a).[25] This effect results from increase of the headspace capacity, as discussed in Section 4.8. A few minutes extraction equilibration time indicates that the majority of analytes extracted by the fiber originate from the headspace. This conclusion is further supported by the stirring data obtained for 60°C (Figure 4.20b). Already at this temperature, the amount of analyte recovered does not depend very much on the agitation conditions of the aqueous matrix. The extracted amount is decreased compared to room temperature, but it provides sufficient sensitivity to quantify methamphetamine at ppb levels adequate for screening applications.

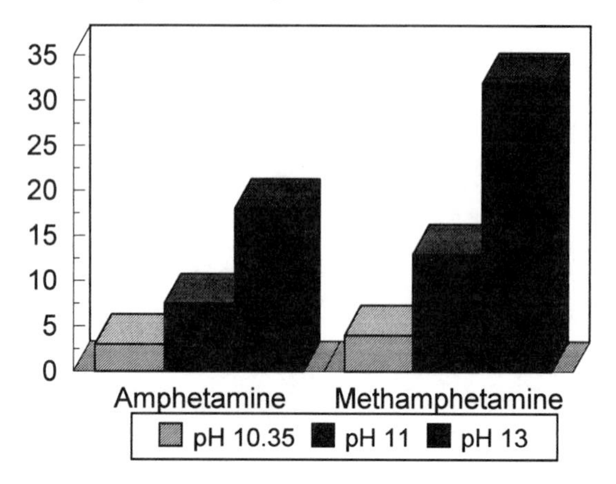

Figure 4.21 Effect of pH on amphetamine and methamphetamine extraction.

pH Adjustment Can Improve Sensitivity for Basic and Acidic Analytes. Sample pH affects the dissociation equilibria in aqueous media. For example, a decrease in pH results in concentration increase of neutral species of acidic compounds present in the sample and, therefore, increase in the extracted amount. For full conversion of species to neutral forms, the pH should be at least two units below the pK of a given analyte. For the basic analytes, the pH must be larger than pK + 2 (see Chapter 3). Figure 4.21 illustrates the increase in the amount of amphetamines and metaamphetamines extracted as pH is increased.[26] The pH optimization should include experimental verification of the expected results since adding a buffer to the sample modifies the matrix, which results in distribution constant changes. In practice, it is very difficult to implement pH change with the direct extraction approach since contact of the fiber with high and low pHs damages the coating. However, headspace SPME is a natural choice for use with a pH-modified matrix.

The Sample Should Be Buffered to Ensure Good Reproducibility When Basic or Acidic Compounds are Present in the Sample. Samples may vary substantially in pH when they are obtained from the source and delivered to the laboratory. To ensure good reproducibility, variation in pH levels should be eliminated by adding small amounts of an appropriate buffer.

Addition of Salt to Aqueous Samples Generally Increases the Fiber/Matrix Distribution Constant of Neutral Organic Molecules. As expected from theory, increases in the amount extracted are observed frequently when the salt concentration is increased. Sometimes, however, when analytes are in dissociated form, a decrease in the amount extracted is observed. This is

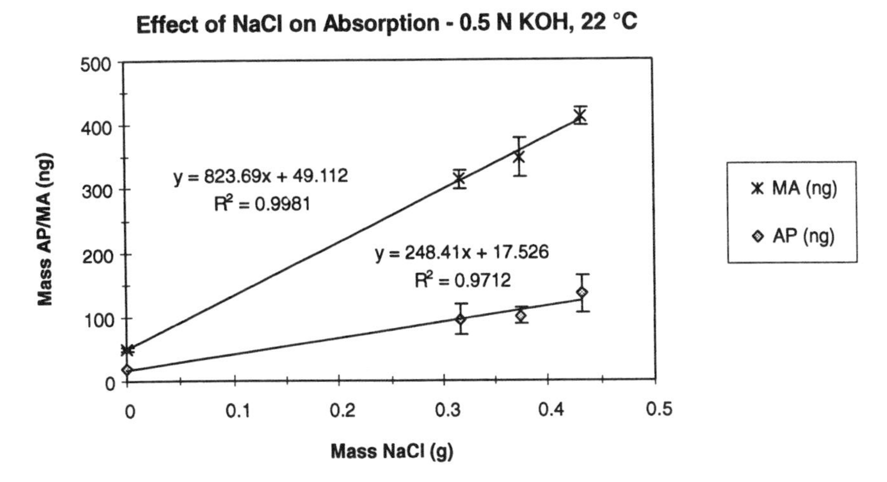

Figure 4.22 Effect of salt concentration on amphetamine (AP) and methamphetamine (MA) extraction.

expected, since the activity coefficient of the ionic species in water increases with the increase of a sample's ionic strength. Therefore, it is important first to convert the analytes into neutral forms. Figure 4.22 illustrates increase of amphetamines extracted at high salt concentration after KOH is added to the solution.

4.13 Determination of Linear Dynamic Range of the Method for a Pure Matrix at Optimum Extraction Conditions

Modification of the extraction condition affects both sensitivity and equilibration time. It is advisable to check previously determined extraction times before proceeding to the determination of the linear dynamic range. This step is required if a substantial change of sensitivity occurs during the optimization. In practice, it is unnecessary to generate a whole new extraction time profile, but rather to check the longest time at which the extracted amount is decreasing.

SPME typically has a very wide linear dynamic range. At this stage, all important extraction parameters have been considered and it is time to check if the linear range of the method consisting of sample preparation, separation and quantitation techniques, covers the target range of the analyte concentration. This measurement will also allow estimation of detection limits. SPME coatings include polymeric liquids such as PDMS, which are used as GC stationary phases, which by definition have a very large linear range. For solid

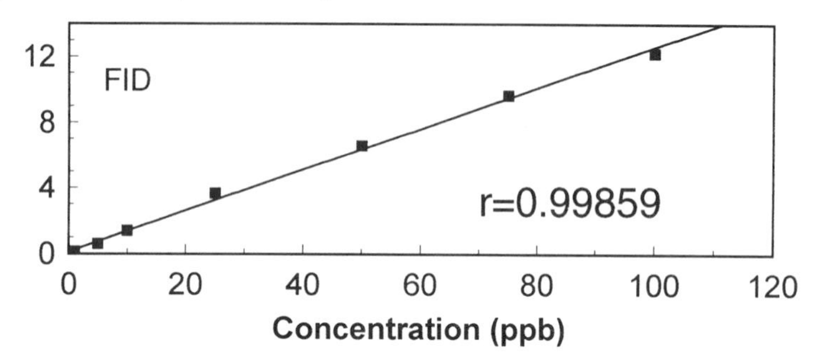

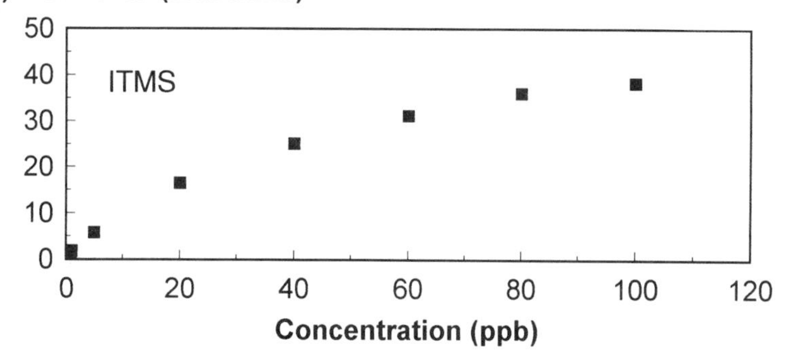

Figure 4.23 Effect of the detector on the method linear dynamic range. Sample: aqueous solution of tetraethyllead. Fiber coating: 100 μm PDMS.

sorbents, such as Carbowax/DVB, or PDMS/DVB, the linear range is smaller because of a limited number of sorption sites on the surface, but it still has a linear range of several orders of magnitude for typical analytes. In some cases, when analytes have an extremely high affinity toward the surface, such as basic proteins adsorption on poly(acrylic acid)[27] or bioaffinity coating, the saturation can occur at a low concentration level. In this case, the linear range can be expanded by reducing the exposure time. In the majority of practical applications developed to date, detector dynamic range limits the linear response of SPME methods. As the example in Figure 4.23 illustrates, a wide linear dynamic range is obtained by the SPME/GC method combined with FID (Figure 4.23a) for the analysis of tetraethyllead in water. However, the dynamic linear range is more limited when an ion trap mass spectrometer is used as a detector (Figure 4.23b). This is caused by a fast decomposition of the primary ions produced from the target analyte in the trap due to secondary reactions. On the other hand, the excellent sensitivity of the MS detector allows low ppt determination.

4.14 Selection of the Calibration Method

The standard procedures used for calibration with other extraction techniques can be adopted by SPME. The fiber blank will verify that neither the instrument nor the needle is contaminated by analyte or interfering compound. Typically, the fiber coating is "cleaned" by desorption in the injector, or the specially designed conditioning device, at the optimized conditions prior to injection as a "blank". This process ensures that the fiber coating itself does not contain contaminants.

For "clean" sample matrices, the external calibration and literature *K* values can be used. When the matrix is simple, such as air or clean water, the distribution constants are very similar to the pure matrix. It has been shown, for example, that the typical moisture level in ambient air, as well as the presence of salt and/or alcohol in water, is less than the 1% level and does not change K beyond 5% RSD, typical for SPME determinations. In many such instances, calibration might not be necessary, since the appropriate distribution constants which define the external calibration curve are available from the literature, or they can be calculated from chromatographic retention times (see Chapter 5).

A special calibration procedure, such as isotopically labeled spikes or standard addition, should be used for more complex samples. In these methods, we assume that the target analyte behaves similarly to spikes during the extraction. This is a good assumption when analysing homogeneous samples. However, this might not be true when heterogeneous samples are analysed, unless the extraction conditions are such that native analytes are quantitatively released from the matrix.

4.15 Optimization of Extraction Conditions for Heterogeneous Samples

Optimizing the process that releases native analytes from the matrix is more difficult since the analyte/matrix association is usually poorly understood. Whenever a new type of complex matrix is considered, a small research project must be conducted to find the optimum extraction conditions which give the fastest and most complete release and partitioning of native analytes. Typically, an empirical approach is taken and several parameters are varied, such as temperature and type of additives. A change of extraction phase type (used in optimization of many solvent extraction methods) is of limited value in SPME since the fiber coating does not interact directly with solids present in the matrix. Heating can release analytes from a solid matrix, but sometimes heating is not adequate because analytes are bound too strongly or the matrix or analytes are unstable at higher temperatures. Additives (small amounts of appropriate compounds added to the matrix or extracting phase to enhance

Table 4.8 Quantitation of PAHs from Urban Air Particulate (NIST 1649) by Static High Temperature Water Extraction

PAH	Cert Conc [μg/g(%RSD)]	Estimated Concentration as % of Certified Concentration (% RSD)		
		For 100 mg, 270°C	For 50 mg, 270°C	For 50 mg, 300°C
Fluoranthene	7.1 (7)	137 (6)	149 (8)	128 (2)
Pyrene	7.2 (7)	113 (9)	121 (7)	105 (7)
Benzo(*a*)pyrene	2.9 (17)	49 (14)	72 (10)	70 (4)

the extraction of analytes) are an alternative to heating. The additive should be selected to enhance the release of analytes from the matrix, but not interfere with the partitioning of analytes into the fiber coating or with the analysis of the extracted complex mixture. Successful modifiers for SPME have included nonvolatile acids and salts, and water.[14,28] Better understanding of analyte-matrix interaction would allow a more rational choice of appropriate additives. In-depth investigation of sample matrices should be considered and is strongly encouraged. Such information would allow a more logical choice of extraction parameters based on models, when the characteristics of the sample and the matrix are known.

Frequently the optimum conditions for releasing analytes from the matrix are different from the optimum conditions for partitioning them into the coating. In most cases, a high temperature is used to remove native analytes from the matrix. Higher temperatures, however, reduce the distribution constant and sensitivity of the method. To eliminate this disadvantage, the cooled fiber approach could be applied to condense analytes onto the fiber.[29]

Optimization of extraction parameters is considered to be complete when the recoveries of analytes present in a native sample match those introduced with a spike. This verification is performed based on extraction results obtained for certified standard reference materials or by comparing with standard extraction methods. When this condition is accomplished, quantitation is performed by comparing extracted amounts of native and spiked analytes. Table 4.8 illustrates the comparison of SPME results with certified values obtained for Urban Air Particulate (NIST 1649).[30] Exposure to static hot water at 270°C was followed by a cooldown period to increase the amount of analyte absorbed by the PDMS coating.

4.16 Verification of Equilibration Time, Sensitivity, and Linear Dynamic Range for Complex Sample Matrices

The presence of a sample matrix can change not only the distribution constant, but also the equilibration time. In cases where the matrix has additional

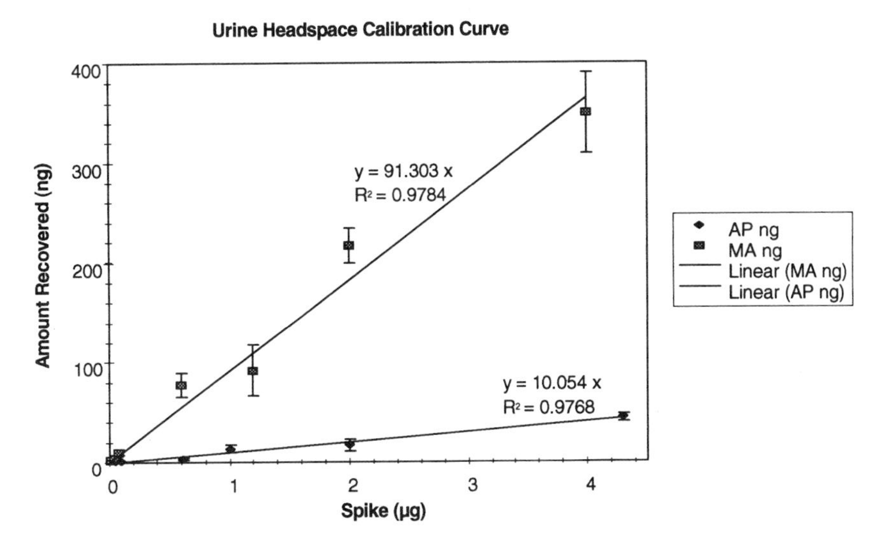

Figure 4.24 Calibration curve for the determination of amphetamine and methamphetamine in urine.

phases, for example, an aqueous sample with solid particulate matter, the kinetics of analyte release might determine the overall extraction rates. To operate at optimum sensitivity, the extraction time needs to be adjusted accordingly.

The distribution constant is expected to change substantially when the coating is swollen by an interfering component in the matrix. To produce a swelling effect, the amount of interference which needs to be absorbed must be a substantial portion of the coating (>1%), which translates for 0.1 mm PDMS to about 10 μg. It is difficult to accomplish this condition in trace analysis since it requires a compound which has good solubility in both the aqueous matrix and the coating. However, this might occur if interference exists as a separate phase, such as an oily suspension or a dispersed hydrophobic humic material.[31] In such cases, SPME is of limited value as a quantitation tool since the distribution constant and the volume of the coating is expected to change during extraction. It should be kept in mind that solid sorbents are much more effected by interferences compared to liquid coatings.

It is helpful to confirm experimentally the linear dynamic range with real matrices. This ensures that the separation conditions are satisfactory and no interference coelute with target analyte. Figure 4.24 illustrates the linear dynamic range obtained for amphetamines in urine samples.

4.17 Method Precision

Table 4.9 summarizes the most important factors controlling precision in methods involving SPME. The majority of the factors have been discussed in the foregoing sections.

Table 4.9 Factors Controlling Precision in SPME Methods

Volume of the fiber coating (thickness and length)
Condition of the fiber coating (cracks, adsorption by high M.W. species)
Moisture in the needle
Temperature
Sample matrix components (salt, organic material, humidity, etc.)
Agitation level
Sampling time, if nonequilibrium conditions are used
Sample volume
Headspace volume
Vial shape
Time between extraction and analysis
Adsorption on the vessel wall
Geometry of the injector
Fiber positioning during injection
Condition of the injector (pieces of septum)
Stability of the detector response

Condition of the Fiber is Very Important. The presence of adsorbed high molecular species, such as proteins and humic matter, on the surface of the fiber causes a change in sorption characteristics. In some cases, the adsorption process can be reversed by soaking the fiber in an appropriate desorption solution. Badly cracked coatings used in direct extraction can transfer a small portion of the sample matrix into the analytical instrument, resulting in its poor operation. The volume of the coating determines the amount of analytes extracted and transferred for analysis. The fibers which begin to loose phase should be replaced immediately. Coating thickness as well as length should be monitored to ensure equivalency when changing the fiber. If the volume of the coating varies, an appropriate correction factor should be incorporated.

Reproducibility of the Sample and Headspace Volumes can Impact the Precision of SPME Methods. From the discussion in Section 4.8, it is evident that the volume of the sample has an impact on the amount extracted when only a few mL sample is used. Therefore, small samples, such as 1 mL, must be measured very carefully. For large samples, a small variation in sample volume will not have a substantial effect on precision of the data. Headspace volume can be an important factor determining the precision of results in three-phase systems. It is relatively easy to measure sample volume accurately. However, vials are not manufactured to have exactly the same volume. Wall thickness and bottom shapes may differ from vial to vial. Also, the shape of the septum in a closed vial can vary from concave to convex. All these factors will affect the total volume of the system, and therefore the headspace volume, which is the difference between total volume and sample volume. Caution needs to be taken to ensure the use of the same total volume vials, so that headspace volume is constant from vial to vial. Such differences (especially related to septum shape) are usually more pronounced in small vials, where a small absolute variation could correspond to a significant fraction of the total volume. Therefore, less precision can usually be expected with their use.

Time Between Preparation of Standards or Sample Collection and Extraction Should Be Minimized. After placing aqueous samples or standards in the vial, analytes begin to interact with the wall of the container. Very hydrophobic substances are frequently strongly adsorbed on the surface and are not available for extraction by the coating. This results in lower than expected extraction amounts, causing a bias toward lower concentration. One way to prevent this is to use vials with more inert surfaces, for example, silanized vials. The other approach is to add a few percent of methanol to the sample matrix.[32] However, presence of methanol in the matrix might result in lowering the distribution constant and sensitivity of the determination (see section 3.2.3).

Time Between Extraction and Analysis Should Be Minimized. When time between the end of extraction and the beginning of analysis is longer that a few seconds, the fiber should be sealed and preferably cooled (dry ice is recommended) to reduce losses and interferences. After the fiber is withdrawn from the sample, it begins to equilibrate with the ambient air. This process is slowed substentially, when the fiber is withdrawn into the needle. Therefore, the few seconds required to inject the sample do not result in any substantial loss of analytes. However, for longer times, sealing of the fiber is necessary. This can be accomplished using a piece of GC septum. In addition, this protection prevents contamination of the fiber with interferences, or even with target analyte which may be present in the ambient air. Cooling of the SPME device might be necessary to extend the stability of the volatile analytes in the coating.[24]

4.18 Method Detection Limits

There are several methods described in the literature to determine the method detection limits. The most widely accepted definition is based on estimating the detection limit using low concentration spikes and calculating the standard deviation of the determination. The detection limit is then defined as 3 times the standard deviation obtained for the analyte concentration not higher than 10 times the method detection limit.

4.19 Validation

Validation of the method might include comparison of quantitation results with certified values obtained for the standard reference materials which have similar matrix and target analytes. The other approach is to validate the method against well-accepted extraction techniques for analysis of target sam-

Table 4.10 Regression Line Parameters for SPME Versus Purge-and-Trap for Clean Water Analysis*

Compound	Regression Parameters: Purge-and-Trap on x Axis		
	Slope	y Intercept	Correlation
Benzene	1.07 ± 0.087	−5.9 ± 6.6	1
Toluene	1.06 ± 0.06	−4.6 ± 4.6	1
Ethylbenzene	1.06 ± 0.046	−3.8 ± 3.7	1
m,p-Xylene	1.05 ± 0.073	−8.4 ± 11	1
o-Xylene	1.07 ± 0.060	−4.0 ± 4.6	1

* Source: Ref. 33.

ples and analytes. Table 4.10 summarizes results of multilevel validation work obtained when the standard purge-and-trap technique is compared with SPME for analysis of BTEX compounds in water. The regression line has a slope close to 1 with a very small intercept value and better than 0.99 correlation for all species. The very good agreement between the two methods indicates suitability of headspace SPME for volatile organic compound analysis in aqueous matrices.[33]

Finally, interlaboratory studies are frequently performed to confirm that SPME can be used successfully in different environments. Table 4.11 summa-

Table 4.11 Statistical Characteristics of the Results Obtained by the Participating Laboratories for the Blind Sample in the Round Robin Test

Compound	Statistical Characteristics[a]						Confidence Intervals (μg/L)[b]	
	s_r	s_L	s_R	r	R	G.A.	C.I.	T.V.
Dichlorvos	2.06	5.04	5.44	5.83	15.4	27.3	27 ± 5.8	25 ± 1.35
EPTC	0.56	1.56	1.66	1.57	4.7	9.9	10 ± 1.6	10 ± 0.54
Ethoprofos	0.82	4.79	4.86	2.32	13.74	15.5	16 ± 2.3	17 ± 0.92
Trifluralin	0.27	0.57	0.63	0.76	1.79	1.6	1.6 ± 0.76	2 ± 0.11
Simazine	2.34	3.45	4.17	6.61	11.79	23.6	24 ± 6.6	25 ± 1.35
Propazine	1.21	2.04	2.37	3.42	6.71	9.5	10 ± 3.4	10 ± 0.54
Diazinon	0.63	2.13	2.22	1.79	6.29	8.2	8 ± 1.8	10 ± 0.54
Meta-chlorpyriphos	0.12	0.32	0.34	0.35	0.97	1.6	1.6 ± 0.35	2 ± 0.11
Heptachlor	2.03	2.89	3.53	5.75	10	8.9	9 ± 5.8	10 ± 0.54
Aldrin	0.54	0.73	0.91	1.53	2.58	2.0	2 ± 1.5	2 ± 0.11
Metolachlor	0.73	2.83	2.92	2.07	8.28	15.7	16 ± 2.1	17 ± 0.92
Endrin	0.87	3	3.13	2.47	8.85	8.8	9 ± 2.5	10 ± 0.54

[a] s_r, repeatability standard deviation; S_L, inter-laboratory standard deviation; s_R, reproducibility standard deviation; r, repeatability; R, reproducibility; G.A., gross average.
[b] of the gross average and "true" value, respectively.

rizes the statistical characteristics of the results corresponding to a blind sample test performed on the SPME pesticide method which involved several laboratories in Europe and North America. The results indicate that the performance of the method was reproduced in different laboratories.[34] In general, the results are characterized by good repeatability, which proves that SPME is a valid method for the determination of pesticides at trace levels, a very diverse group of semivolatile compounds. As expected, the interlaboratory and reproducibility standard deviations are higher, since they consider differences between laboratories. However, they are still satisfactory (generally $<5\%$).

The results presented in Table 4.11 indicate that SPME is an accurate method. In all cases, the confidence intervals of the gross average and the "true" value overlap, which indicates that any differences between the two respective values are due to random factors. Interestingly, for 10 out of 12 compounds, the values of the gross average are slightly lower than the "true" values. This might be due in part to the losses of analytes through adsorption (as described earlier) in cases where the aqueous solutions were not prepared immediately before the analysis.

4.20 Automation

SPME is a very powerful investigative tool, but it can also be a technique of choice in many applications for processing a large number of samples. To accomplish this task would require automation of the developed methods. As SPME automated devices with more advanced features and capabilities become available, automation of the developed methods becomes easier. In some cases, custom-made modifications to commercially available systems can facilitate operation of a given method closer to optimum conditions.

References

1. V. Mani (Supelco), private communication.
2. Z. Zhang and J. Pawliszyn, *J. Phys. Chem.*, ***100***, 17648 (1996).
3. P. Martos, A. Saraullo, and J. Pawliszyn, *Anal. Chem.*, ***69***, 300 (1997).
4. A. Saraullo, P. Martos, and J. Pawliszyn, *Anal. Chem.*, in press.
5. D. Potter and J. Pawliszyn, *Environ. Sci. Technol.* ***28***, 298 (1994).
6. K. Buchholz and J. Pawliszyn, *Environ. Sci. Technol.* ***27***, 2848 (1993).
7. A. Boyd-Boland, S. Magdic, and J. Pawliszyn, *Analyst* ***121***, 929 (1996).
8. T. Gorecki, P. Martos, and J. Pawliszyn, *Anal. Chem.*, submitted.
9. E. Fogelqvis, B. Josefsson, and C. Roos, *High Resolut. Chromatogr.* ***3***, 568 (1980).
10. L. Pan and J. Pawliszyn, *Anal. Chem.*, ***69***, 196 (1997).
11. Y. Cai and J. Bayona, *J. Chromatogr.* ***696***, 113 (1995).

12. T. Gorecki and J. Pawliszyn, *Anal. Chem.*, ***68***, 3008 (1996).

13. L. Pan, M. Adams, and J. Pawliszyn, *Anal. Chem.* ***67***, 4396 (1995).

14. Z. Zhang and J. Pawliszyn, *Anal. Chem.* ***67***, 34 (1995).

15. S. Motlagh and J. Pawliszyn, *Anal. Chim. Acta* ***284***, 265 (1993).

16. R. Eisert and K. Levsen, *J. Chromatogr.* ***737***, 59 (1996).

17. Z. Zhang, T. Gorecki, and J. Pawliszyn, in preparation.

18. R. Shirey, *High Resolut. Chromatogr.* ***18***, 495 (1995).

19. T. Gorecki and J. Pawliszyn, *Anal. Chem.* ***67***, 3265 (1995).

20. J. Chen and J. Pawliszyn, *Anal. Chem.* ***67***, 2530 (1995).

21. H. Diamon and J. Pawliszyn, in preparation.

22. T. Gorecki and J. Pawliszyn, *Analyst*, submitted.

23. J. Bourke, T. Spittler, and S. Young, in *Principles of Environmental Sampling*, L.H. Keith (Editor), Washington, DC, 1988), pp. 359.

24. M. Chai and J. Pawliszyn, *Environ. Sci. Technol.* ***29***, 693 (1995).

25. H. Lord and J. Pawliszyn, *Anal. Chem.*, submitted.

26. L. Wang, *Determination of Amphetamine and Methamphetamine in Urine and Blood Plasma by Headspace Solid Phase Microextraction*, M.Sc. Thesis (University of Waterloo, Waterloo, Canada, 1996).

27. J-L. Liao, C-M Zeng, S. Hjerten, and J. Pawliszyn, *J. Microcolumn Sep.* ***8***, 1 (1996).

28. K. Buchholz and J. Pawliszyn, *Anal. Chem.* ***66***, 160 (1994).

29. Z. Zhang and J. Pawliszyn, *Anal. Chem.* ***67***, 34 (1995).

30. H. Diamon and J. Pawliszyn, *Anal. Commun.* ***33***, 421 (1996).

31. Z. Zhang, J. Poerschmann, and J. Pawliszyn, *Anal. Commun.* ***33***, 219 (1996).

32. J-P. Beaudonnet, private communication.

33. B. MacGillivray, P. Fowlie, C. Sagara, and J. Pawliszyn, *J. Chromatogr. Sci.* ***32***, 317 (1994).

34. T. Gorecki and J. Pawliszyn, *Analyst*, ***121***, 1381 (1996).

CHAPTER

5

Applications

SPME methods are still at the initial stages of development. Early interest in this technology is driven by laboratory applications, which explore its solvent-free feature, speed of extraction and convenient automation and hyphenation with analytical instruments. Some other unique opportunities, such as direct field measurements, investigation of multiphase equilibrium processes and analysis of large biomolecules are still to be investigated. The majority of the methods have been developed for clean matrix samples, and only a few optimization parameters have been fully explored. However, the results obtained are very promising, giving ideas about the potential scope of successful applications. A good understanding is emerging regarding approaches which can be taken to the basic matrix types: gasses, liquids, and solids. Progress in this direction is summarized in the first three parts of this chapter. General approaches to facilitate successful extraction of these systems are emphasized. Future research on SPME methods for these matrices is discussed. In the following parts of the chapter, typical applications for various areas of analytical chemistry, are described followed by anticipated future directions. In the last section, physical chemical applications are described with an emphasis on investigations, leading to a better understanding of the extraction process and its potential. It should be noted that the objective of this chapter is not to demonstrate a comprehensive review, but to give only examples of successful SPME applications for different areas. Most examples are based on the author's research experience. A more complete list of developed applications is provided at the end of the book in Appendix B.

5.1 Gaseous Matrix

All SPME methods developed to date for gas analysis are directed toward air applications. However, the approaches described are suitable for analysis of other gas mixtures pending the development of appropriate coatings. Currently, a number of commercially available approaches can be used for the determination of trace contaminants in air. These include grab sampling with stainless steel canisters or nylon bags followed by concentration over a sorbent bed, direct concentration over sorbent using portable pumps, and passive diffusion monitors and others. These techniques are expensive and time consuming to operate. They require the use of adsorbents such as charcoal, silica, or other polymers, whose breakthrough volumes are strongly affected by humidity. The adsorption of contaminants is followed by a desorption step, principally using solvents, and the introduction to an analytical instrument for analysis.[1]

SPME has the potential to dramatically improve cost efficiency of air analysis, since it can integrate the first steps of the analytical process: sampling, extraction and concentration, and convenient introduction to the analytical instrument. Model studies of air extraction have been performed by using two methods, the static and the more realistic dynamic approaches. In the static method, a glass bulb can be applied as shown on Figure 5.1. In this approach, the target compound is introduced to the sealed bulb. After the analytes have been dissolved in air, the fiber is introduced to the bulb through the septum.[2] In practical ambient air measurements, the system is not static, convection is present. Therefore, it is more appropriate to use dynamic flowing gas chambers for the modeling studies. Figure 5.2 shows an example of a simple gas chamber which can be constructed by using standard components.[3]

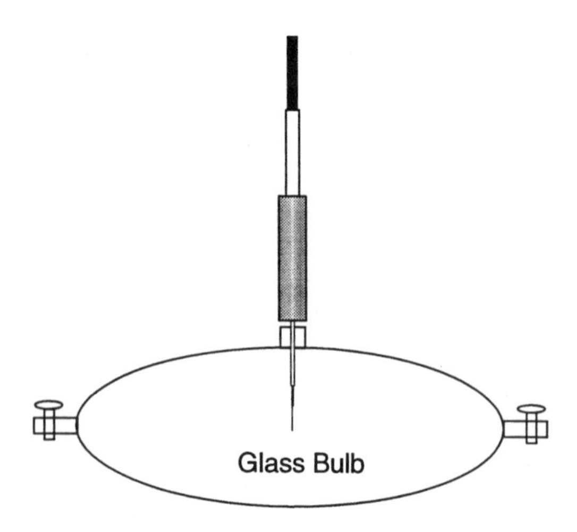

Figure 5.1 Gas sampling by SPME from a bulb.

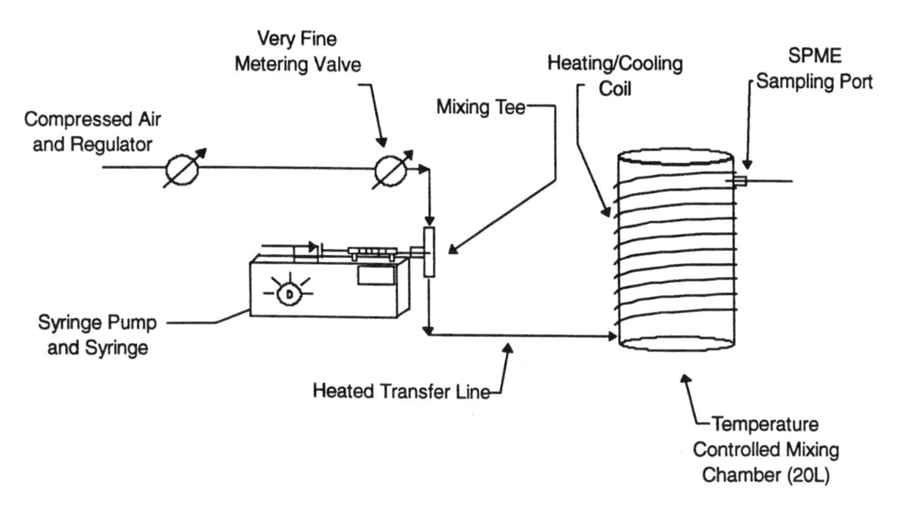

Figure 5.2 Schematic of the standard gas generating chamber. Humidification system not shown.

Equilibration times for the extraction of trace contaminants from moving air are fast, as predicted from theory. Figure 5.3 shows the equilibration time profiles obtained for a 20 L chamber operated at 200 mL/min. The equilibration times are between 20 and 100 seconds, for compounds ranging from benzene and 1,3,5-trimethyl benzene, respectively. They are close to theoreti-

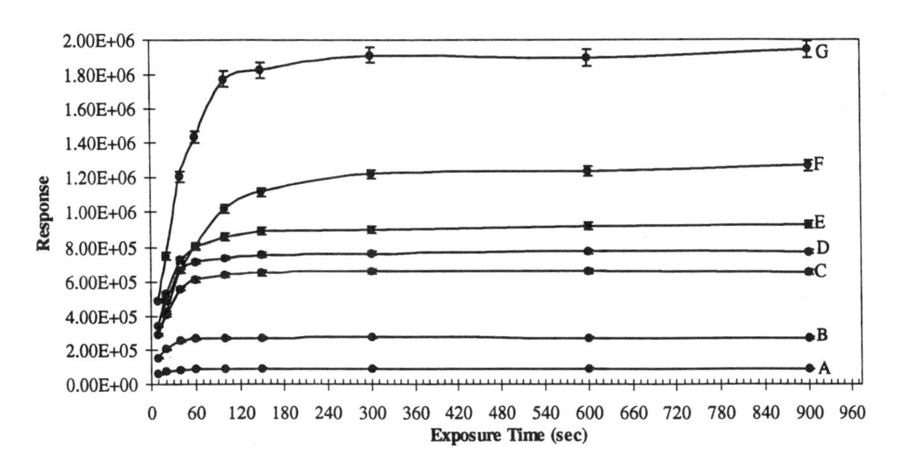

Figure 5.3 Representative absorption/time profiles obtained for extraction with PDMS fiber of a standard gas mixture containing (A) benzene, (B) toluene, (C) ethylbenzene, (D) *p*-xylene, (E) *o*-xylene, (F) α-pinene and (G) 1,3,5-trimethylbenzene.

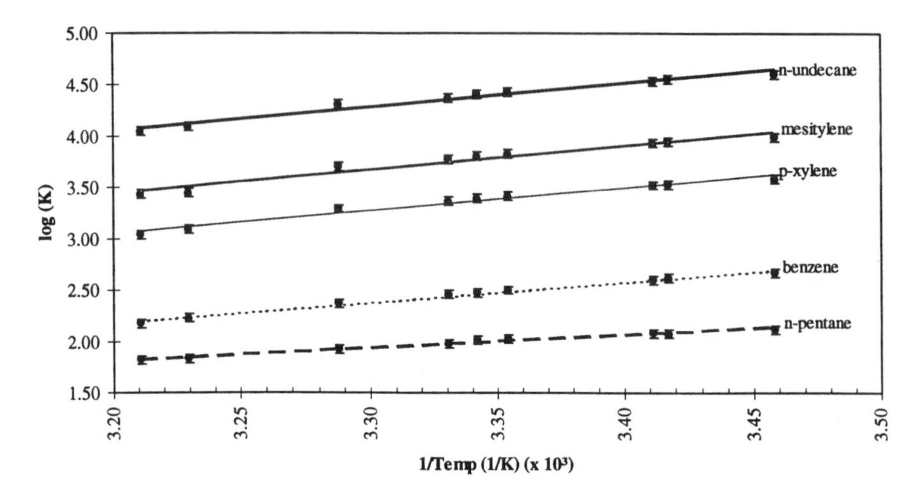

Figure 5.4 Representative plots for pentane, benzene, *p*-xylene, mesitylene and *n*-undecane illustrating the linear relationship between log K_{fg} and $1/T$.

cal values for perfect agitation conditions (see chapter 3). Increased airflow will decrease the equilibration times only for less volatile analytes.

Air is a simple matrix for sampling by SPME. A fiber coating can be selected not to concentrate major air components such as nitrogen, oxygen, carbon dioxide and moisture, but to selectively extract dissolved organics. PDMS has properties close to these characteristics. Only at humidities approaching 100% does it absorbs sufficient amounts of moisture to cause a change in the coating polarity resulting in a small change in the distribution constant. This effect can be conveniently compensated for, if necessary, by measuring the relative humidity of the sample and adjusting the response appropriately. The experimental parameter primarily affecting response is temperature. However the effect of the temperature change on the distribution constant can be conveniently predicted since log K_{fg} is linearly related to $1/T$ and the heat of vaporization of the pure solute, ΔH^V:

$$\log_{10} K_{fg} = \frac{\Delta H^V}{2.303RT} + \left[\log_{10}\left(\frac{RT}{\gamma_i p^*}\right) - \frac{\Delta H^V}{2.303RT^*}\right] \tag{5.1}$$

where p^* is the analyte vapor pressure at a known temperature T^* for a pure solute and γ_i is the activity coefficient of the solute in the coating. In other words, K_{fg} can be calculated for given extraction conditions by measuring the temperature of the sample and knowing the heat of vaporization for the target compound. Figure 5.4 illustrates the above relationship graphically for a range of compounds varying in volatility. The linear relationship is clearly illustrated. In addition, the heat of vaporization and the activity coefficient are related to the retention time of the compound on the GC column using the same

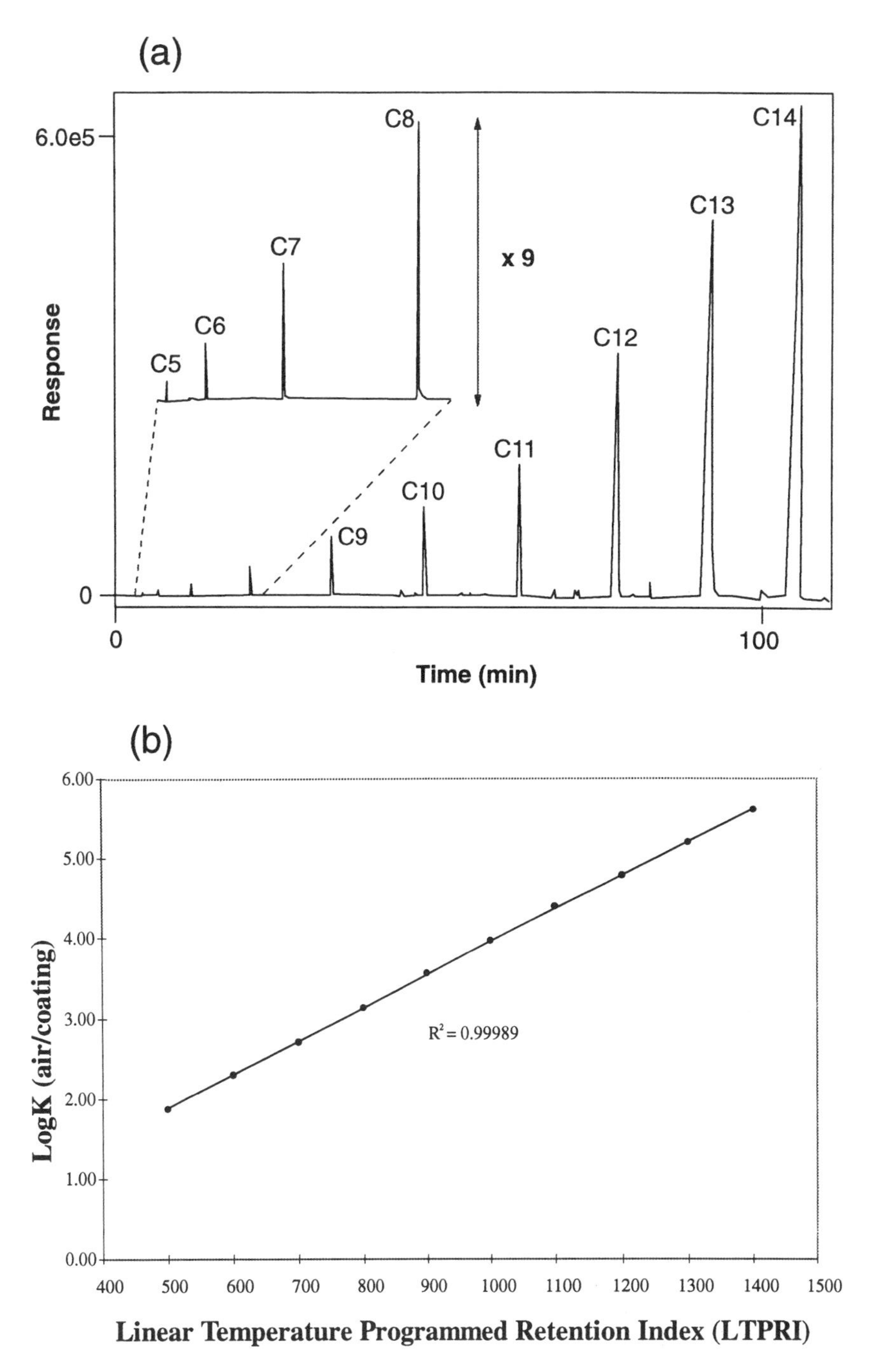

Figure 5.5 Correlation between analyte retention times and K_{fg}. (a) Chromatogram of *n*-alkanes on PDMS column extracted by PDMS coated fiber from air containing 15 ng/mL of each hydrocarbon. Temperature programming conditions: 35°C, 1 deg/min to 220°C. (b) Log K_{fg} (PDMS) as a function of LTPRI for *n*-alkanes at 25°C.

Table 5.1 Summary of K_{fg} SPME/PDMS Data for Hydrocarbons at 25°C

Compound	K_{fg}	
	For LTPRI	For SPME[a]
3-Methylpentane	157	159
2,4-Dimethylpentane	246	262
2,2,3-Trimethylbutane	258	280
2-Methylhexane	351	387
2,3-Dimethylpentane	358	412
2,2-Dimethylhexane	616	673
2,5-Dimethylhexane	672	587
2,2,3-Trimethylpentane	672	569
2,3-Dimethylhexane	877	968
2-Methylheptane	933	993
4-Methylheptane	947	1,060
3-Methylheptane	1,010	1,090
3-Ethylhexane	1,020	990
2,5-Dimethylheptane	1,880	1,970
3,5-Dimethylheptane (D)	1,880	1,960
3,3-Dimethylheptane	1,900	2,090
3,5-Dimethylheptane (L)	1,900	2,100
2,3-Dimethylheptane	2,270	2,390
3,4-Dimethylheptane (D)	2,310	2,420
3,4-Dimethylheptane (L)	2,330	2,620
2-Methyloctane	2,490	2,600
3-Methyloctane	2,670	2,890
3,3-Diethylpentane	2,690	2,610
2,2-Dimethyloctane	4,240	4,320
3,3-Dimethyloctane	4,990	5,050
2,3-Dimethyloctane	6,020	6,100
2-Methylnonane	6,690	6,690
3-Ethyloctane	6,890	6,970
3-Methylnonane	7,150	7,100
Benzene	296	301
Toluene	815	818
Ethylbenzene	2,020	2,070
m-Xylene	2,190	2,090
p-Xylene	2,220	2,500
o-Xylene	2,710	2,900
Isopropylbenzene	3,780	3,880
n-Propylbenzene	4,960	5,040
1-Methyl-3-Ethylbenzene	5,340	4,750
1-Methyl-4-Ethylbenzene	5,440	6,230
1,3,5-Trimethylbenzene	6,150	6,480
1-Methyl-2-Ethylbenzene	6,260	6,580
Isobutylbenzene	8,450	8,360
sec-Butylbenzene	8,680	8,590
1-Methyl-3-Isopropylbenzene	9,660	10,100
1-Methyl-4-Isopropylbenzene	9,920	10,200
1-Methyl-2-Isopropylbenzene	11,200	12,000

(Continued)

Table 5.1 *(Continued)*

Compound	K_{fg}	
	For LTPRI	For SPME[a]
1-Methyl-3-n-Propylbenzene	12,800	13,200
1,3-Dimethyl-5-Ethylbenzene	13,700	15,000
1-Methyl-2-n-Propylbenzene	14,700	14,900
1,4-Dimethyl-2-Ethylbenzene	16,200	15,900
1,2-Dimethyl-4-Ethylbenzene	17,400	17,400
1,3-Dimethyl-2-Ethylbenzene	18,300	18,100
1,2-Dimethyl-3-Ethylbenzene	20,900	20,000
1,2,4,5-Tetramethylbenzene	23,400	24,700
2-Methylbutylbenzene	24,000	24,100
tert-1-Butyl-2-Methylbenzene	27,100	26,200
n-Pentylbenzene	35,300	34,500
t-1-Butyl-3,5-Dimethylbenzene	43,900	45,600
t-1-Butyl-4-Ethylbenzene	44,900	43,700
1,3,5-Triethylbenzene	66,800	67,300
1,2,4-Triethylbenzene	77,600	75,600
n-Hexylbenzene	95,700	90,100
Pentane	77	*
Hexane	201	*
Heptane	521	*
Octane	1,356	*
Nonane	3,525	*
Decane	9,166	*
Undecane	23,834	*
Dodecane	61,973	*
Tridecane	161,139	*
Tetradecane	418,986	*

[a] Asterisk indicates compounds used as calibration standard.

coating material.[4] Therefore, retention times of the eluting compound should directly provide the appropriate distribution constants. Figure 5.5a illustrates this statement by showing progressively higher extraction efficiencies of n-alkanes from air with increase of the retention time for given hydrocarbon.

A more universal approach to describe the relationship is to use the Kovats retention indices since they are available in the literature. However, calculation of Kovats indices requires isothermal chromatographic operation. This limits the range of compounds that can be studied in one run. The more appropriate approach is to use the linear temperature programmed retention index (LTPRI) to estimate the retention indices of a wide range of analytes varying in volatility.[5] This scale allows the assignment of K_{fg} values, since a well-defined relationship exists between the index and K_{fg} (see Section 3.2.2). Figure 5.5b shows the linear relationship between the PDMS/gas distribution constant

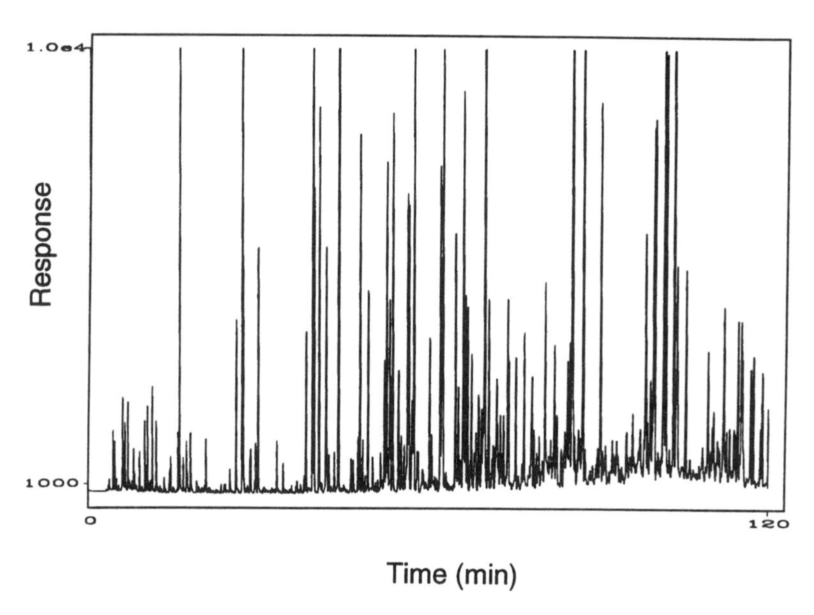

Figure 5.6 GC/FID chromatogram of airborne gasoline extracted with PDMS fiber.

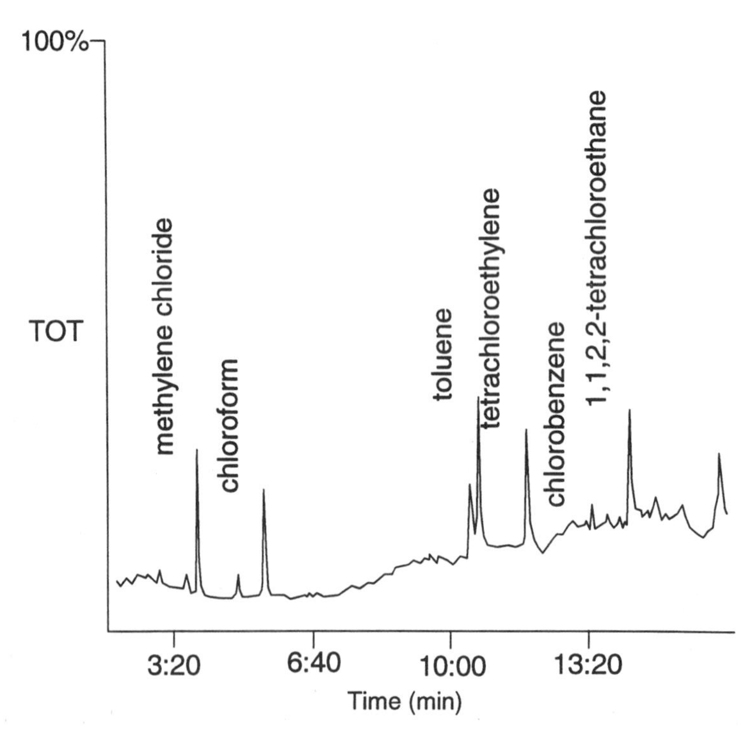

Figure 5.7 Components of synthetic organic laboratory air sample extracted by PDMS fiber.

Source: Adopted with permission from ref. 2.

and LTPRI obtained on the PDMS column for a range of linear hydrocarbons.[6] Coordinates of the line are:

$$\log_{10} K_{fg} = 0.00415 \times LPTRI - 0.188 \quad (5.2)$$

This relationship can be applied to calculate the PDMS/gas distribution constant of any analyte as long as the retention index for a given compound on a PDMS column is available from the literature or by experiment. Table 5.1 summarizes the distribution constant data obtained using this approach, and compares it to the direct SPME experimental values for a range of hydrocarbons. The observed differences are small, indicating that eq. 5.2 can be used to determine PDMS/gas distribution constants. This approach can be extended to other coatings as long as appropriate columns are available. In addition, the compounds do not need to be identified to have appropriate distribution constants assigned to them. This approach allows quantitation without identification and can be applied, for example, to determining the total petroleum hydrocarbon (TPH) content of air samples.

For example, Figure 5.6 shows the gasoline range hydrocarbons extracted from the air in the vicinity of a fuel pump. FID produces a signal proportional to the number of carbons reaching the detector which translates to the mass of analyte extracted by the fiber and transferred onto the column. Therefore, it is possible to quantify the amount of each hydrocarbon in the sample from the corresponding peak area and the LTPRI calculated from the retention time. In practice, the extracted mixtures are frequently too complex to be separated into individual components as in Figure 5.6. To obtain the TPH value in this case, the chromatogram can be divided into small retention time windows, followed by the integration of the FID trace over the appropriate ranges. Typically, the calibration of the FID response is performed using a single standard, such as toluene. Determination of hydrocarbon quantity in the air is based on K_{fg} values assigned to a particular range of retention times. The total amount of hydrocarbons is calculated by adding the values corresponding to each window. The TPH results using SPME/PDMS and the standard charcoal tube technique obtained for the complex mixture defined by chromatogram in Figure 5.6 are in 262 and 247 μg/L, respectively.[6] These values show very good agreement, indicating that the approach just described can be used to quantify analytes in the gaseous samples without need for identification.

SPME has superior sensitivity for short term monitoring compared to traditional devices, which are limited by the gas throughput.[3] Figure 5.7 shows the chromatogram obtained after SPME air sampling in a synthetic organic laboratory. Several solvents, which had been used a few days previously by students, were detected in the labratory air. The SPME fibers are designed to work at high temperatures. For example, Figure 5.8 illustrates the identification of several PAHs in diesel exhaust, which indicate the typical distribution of these contaminants, as previously determined by more expensive methods. The sensitivity of the SPME technique can be further improved by using

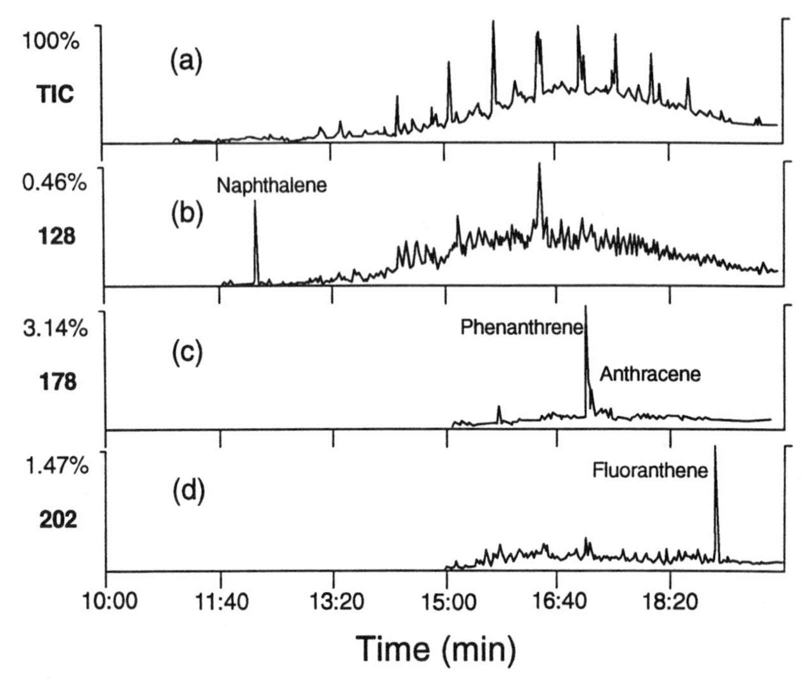

Figure 5.8 GC/MS traces corresponding to polycyclic aromatic hydrocarbons present in diesel exhaust sampled by SPME/PDMS; (a) total ion chromatogram; selected ion traces at: (b) m/z = 128, (c) 178, (d) 202.

Source: Adopted with permission from ref. 2.

thicker, more selective coatings, or by cooling the fiber to increase the PDMS/gas distribution constant. The other approach is to incorporate the derivatization reagent in the coating to allow "trapping" of the analyte in the coating or to use carbon based coatings. All these modifications will result in SPME sensitivity enhancement for air monitoring.

The advantages of SPME for long term sampling are less clear at present. Typically, the fiber reaches equilibrium with the air components in the first few minutes and then does not accumulate any more analytes, independent of the exposure time. However, SPME assembly can also work as an integrated sampling device. The simplest way to accomplish this task is to retract the fiber into the needle. In this position, the fiber is surrounded by the needle, resulting in a very slow extraction since the analytes need to first diffuse through the needle opening before they can reach the fiber. The depth of the fiber retraction into the needle defines the diffusion length and the integrating factor. Figure 5.9 shows the response of the device as a function of the exposure time for three organic compounds. About 30 minutes linear response was observed. Extension of this range can be obtained by retracting the fiber

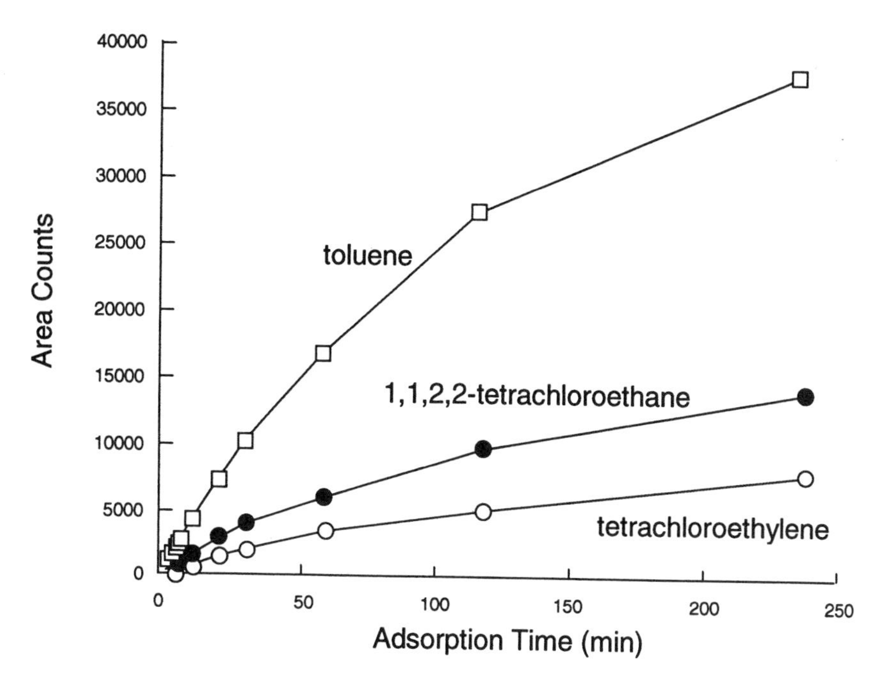

Figure 5.9 Integrated sampling with PDMS fiber retracted inside the needle of an SPME device.

Source: Adopted with permission from ref. 2.

further into the needle, increasing capacity of the sorbing phase or by placing an additional barrier in front the needle opening, for example a membrane, which can increase the selectivity of permeation (see Section 3.3.5). An alternative approach is to use a chemical reaction in the coating.

Figure 5.10 shows the extraction of low molecular carboxylic acids from a small volume of air using a fiber doped with 1-pyrenyldiazomethane.[7] As expected, the linear extracted amount vs time relationship is obtained until about 60% of acid is extracted from the sample. In practice, however, when the fiber is exposed to ambient air, analytes will accumulate linearly with concentration as long as the fiber is exposed or the derivatization reagent is substantially consumed. This approach is particularly useful for integrated sampling of very low analyte concentration in air or other gaseous matrix.

Table 5.2 summarizes a field experiment designed to monitor styrene in industrial air using both a 5-minute grab sample and 30 minutes of integrated measurement. Two standard techniques are compared to SPME with a PDMS fiber exposed for grab sampling and retracted into the needle for integrated sampling. The agreement between the methods is very good considering the preliminary nature of the experiments.[3]

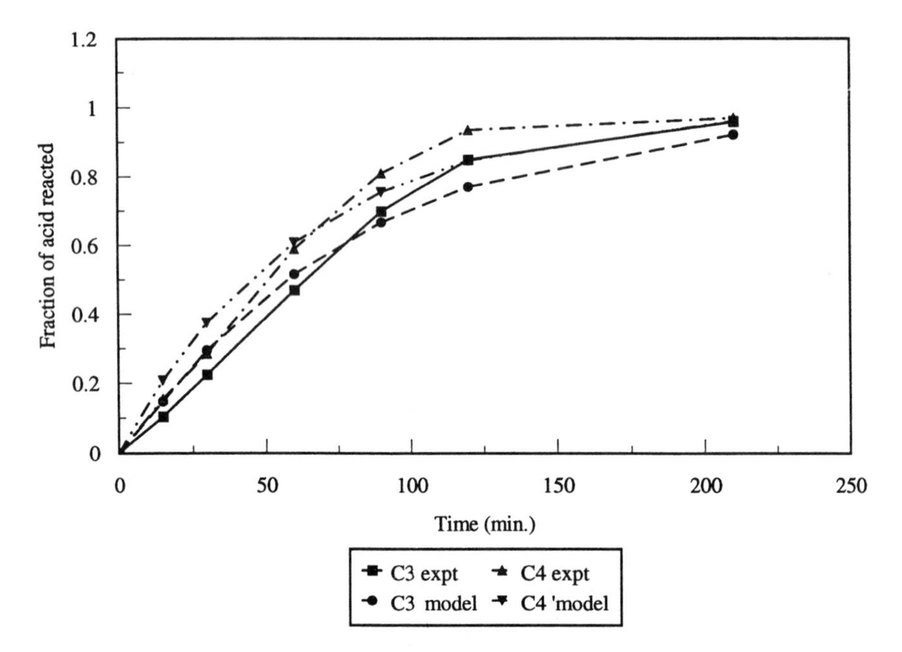

Figure 5.10 Comparison of experimental data versus theoretical model prediction for reaction of gaseous propionic and butyric acids with PDAM in PA-coated fiber.
Source: Adopted with permission from ref. 7.

SPME procedures for analysis of air containing particulates have not yet been fully developed. However, the general direction can be anticipated based on initial data. The direct extraction mode is expected to perform well in most cases. However, for high concentrations of particulate matter, hollow fiber membrane protection might be a more appropriate approach. Concentration of dissolved organics can be quantified by using external calibration, since the total volume of sample is large. The total analyte concentration is typically obtained by adding an internal standard (see Section 5.8.2). However, this

Table 5.2 Comparison of SPME with Other Samplers

	Industrial Concentrations of Styrene (μg/L)		
Sample Type	SPME with PDMS[a]	Charcoal Tube[b]	Passive Badge[c]
Grab (5 min)	130	97	90
Integrated (30 min)	56	54	72

[a] observed concentration styrene with SPME 100 μm PDMS at 296 K and 25% relative humidity for both sampling types.
[b] Active sampling.
[c] Integrated value over the sampling time.

Table 5.3 Effects of Temperature on Sample Storage[a]

	25°C				5°C		−70°C		
	Uncapped		Capped		Capped		Capped		
Compound	2 min	60 min	30 min	60 min	30 min	60 min	60 min	24 h	48 h
Chloroform	97	51	70	62	93	89	95	50	30
1,1,1-Trichloroethane	95	49	67	59	93	91	92	76	45
Carbon tetrachloride	97	54	73	70	95	93	95	79	66
Benzene	94	69	89	81	94	92	97	83	82
Toluene	96	77	86	89	95	95	94	92	85
Tetrachloroethylene	96	79	85	88	95	94	99	97	90
1,1,2,2-Tetrachloroethane	98	70	90	80	94	92	98	95	93

[a] Values in the table correspond to relative amounts of analyte (in %) remaining on the fiber.

Source: Ref. 2.

approach might be difficult to implement practically for field air sampling as it might be impractical to add a standard to the sample. In this case, the total concentration of ambient air can be obtained by using a sorbent filter which would collect both the dissolved organics and particulates. The sorbent can then be extracted by SPME procedures designed for solids or by an alternative method. Combining data obtained for dissolved organics and total amount, the adsorbed quantity of particulates can be calculated.

Frequently, the fibers cannot be immediately analyzed after the extraction, particularly in field analysis since a portable instrument might not be available. In such a situation, the appropriate storage method for the fibers needs to be designed. This can be simply accomplished by retracting the fiber into the needle and sealing its open end with a piece of a septum. Cooling the needle provides additional protection against analyte losses. Table 5.3 summarizes the initial data about the suitability of this simple approach for several solvents varying in volatility. One hour storage with a fiber placed in solid carbon dioxide preserves all analytes well. However, the loss of volatiles is observed for longer times. The loss of analyte can be traced to sorption by the septum, which is typically made of PDMS polymer. For better field performance, the SPME syringe needs to be modified to provide better sealing of the fiber for example, see Figure 2.3B. Also, the current design will benefit from substantial structural changes to facilitate more convenient handling by field sampling staff.

5.2 Liquid Matrix

All methods for liquid matrices reported in the literature, as of the summer 1996 were developed for aqueous samples. However, the technique is not

Table 5.4 The Determination of K_{fw} for PDMS in Water Obtained by the Log K_{fw}-LTPRI Relationship and the Experimentally Established K_{fw} Values Obtained by Direct SPME for 22°C

Hydrocarbon	LTPRI	K_{fw} by LTPRI	K_{fw} by SPME
Benzene	638.6	60	58
Toluene	747.1	184	189
o-Xylene	867.9	565	485

limited to water analysis. As emphasized above, with the application of an appropriate coating, target analytes can be preconcentrated on SPME fibers from other type of matrices. For example, as shown in Chapter 4 (Figure 4.4), Nafion coating can be used to extract polar compounds from fuels. Since most of the discussion in Chapter 4 dealt with optimization of aqueous sample analysis, the discussion in this Chapter will be limited to topics not previously covered.

The fiber coating/water distribution constants can be calculated from the following equation (Chapter 3):

$$K_{fw} = K_{fg} K_{gw} \tag{5.3}$$

where K_{fg} can be calculated from chromatographic data, as discussed above, by using LTPRI. K_{gw} is the gas/water distribution constant for a given analyte and can be found in the Henry's Law constant tables. For example, the equation for a PDMS coating and aqueous matrix can be calculated from the equation below:

$$\log K_{fw} = 0.00415 \times \text{LTPRI} - 0.188 + \log_{10} K_{gw} \tag{5.4}$$

This equation is obtained after substitution of the expression for K_{fg} given in eq. 5.2 (Section 5.1) into eq. 5.3.

Table 5.4 compares values of K_{fw} obtained by using the above equation with experimental values for benzene, toluene, and *o*-xylene. The results agree very well, considering that the errors in determination of Henry's constants are typically above 10%. Therefore, by finding the relationship between K_{fg} and LTPRI for a given coating, the appropriate distribution constant can be conveniently calculated from chromatographic data and literature values of Henry's constants. In addition, the Henry's constants are similar for compounds closely related, as illustrated in Figure 5.11, resulting in a single linear relationship between K_{fw} and LTPRI, characteristic for a group of different types of analytes. As expected, the slope of the curve has a value close to the curve in Figure 5.5b, but the intercept varies and corresponds to a sum of the intercept value from eq. 5.2 plus the average value of the Henry's constant for a given group of analytes. For example, as expected from their high Henry's constant values, K_{fw} values for paraffins are larger compared to aromatic

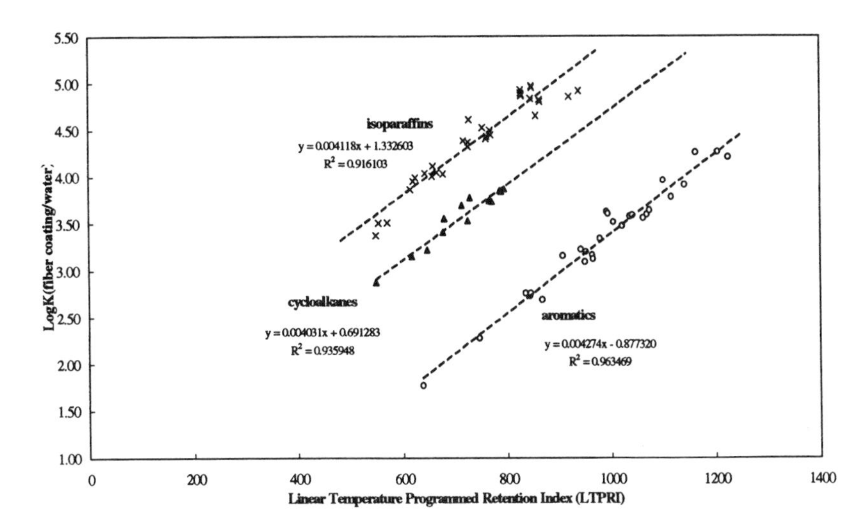

Figure 5.11 Log K_{fw} (PDMS) as a function of LTPRI for isoparaffins, substituted benzenes (aromatics), and cycloalkanes at 25°C.

analytes. Because of this linear relationship, quantitation with minimum identification is possible as long as a detector can selectively assign extracted analytes to appropriate groups of compounds.[8] Table 5.5 summarizes the distribution constants obtained for a number of aromatic and aliphatic hydrocarbons.

Several reports indicated the existence of a linear relationship between the coating/water distribution constant and octanol-water distribution constant, K_{ow}.[9,10] Considering the discussion above, this relationship is expected to exist only for a group of related compounds, such as isoparaffins, cyclohexanes or substituted benzenes. Because the activity coefficients and, therefore, the selectivity corresponding to unrelated groups of analytes in octanol are expected to be different compared to PDMS or other fiber coating, the relationship between the two extracting phases will vary with the change of the chemical properties of analytes. It is possible to design a fiber coating which has a similar structure to octanol and, in this case, the values K_{ow} might be directly applicable. However, for other coatings the trends between K_{fw} and K_{ow} are expected to produce a family of curves as in Figure 5.11. The differences among the curves will correspond to the ratio of appropriate activity coefficients. Therefore, it might be possible to predict the general trends in K_{fw} within a group of related compounds by using corresponding values of K_{ow}.

The above discussion pertains to a pure water sample as a matrix. Presence of other components in water will modify the distribution constants for given analytes. The effect can be predicted based on the discussions in Section 3.2.2. Also, liquid chromatographic experience gives some clues about the trends in the distribution constant change with modification of the matrix. For example, addition of salt would generally result in an increase of the distribution constant

Table 5.5 Summary of K_{fw} Data for Hydrocarbons Obtained by SPME/PDMS at 22°C

Cyclopentane	712
Methylcyclopentane	1,356
Cyclohexane	1,592
cis-1,3-Dimethylcyclopentane	4,289
trans-1,2-Dimethylcyclopentane	3,372
Methylcyclohexane	4,657
cis-trans-cis-1,2,4-Trimethylcyclopentane	5,621
cis-trans-cis-1,2,3-Trimethylcyclopentane	6,556
1-Ethyl-1-methylcyclopentane	6,831
trans-1,2-Dimethylcyclohexane	6,638
cis-cis-cis-1,2,3-Trimethylcyclopentane	7,109
cis-1,2-Dimethylcyclohexane	7,826
2,3-Dimethylbutane	2,359
2-Methylpentane	3,224
3-Methylpentane	3,270
2,2-Dimethylpentane	7,349
2,4-Dimethylpentane	8,989
2,2,3-Trimethylbutane	9,802
3,3-Dimethylpentane	10,963
2-Methylhexane	10,202
2,3-Dimethylpentane	13,074
3-Methylhexane	11,146
3-Ethylpentane	10,816
2,2-Dimethylhexane	24,504
2,5-Dimethylhexane	23,519
2,2,3-Trimethylpentane	21,205
2,4-Dimethylhexane	41,133
2,3-Dimethylhexane	33,749
2-Methylheptane	25,806
4-Methylheptane	27,274
3-Methylheptane	31,856
3-Ethylhexane	28,370
2,5-Dimethylheptane	84,142
3,5-Dimethylheptane (D)	78,829
3,3-Dimethylheptane	76,013
3,5-Dimethylheptane (L)	72,856
2,3-Dimethylheptane	68,675
3,4-Dimethylheptane (D)	93,292
3,4-Dimethylheptane (L)	89,182
2-Methyloctane	45,267
3-Methyloctane	66,682
3,3-Diethylpentane	63,718
2,2-Dimethyloctane	72,155
3,3-Dimethyloctane	82,430
Benzene	58
Toluene	189
Ethylbenzene	566
m-Xylene	533

(Continued)

Table 5.5 *(Continued)*

p-Xylene	564
o-Xylene	485
Isopropylbenzene	1,412
n-Propylbenzene	1,664
1-methyl-3-ethylbenzene	1,231
1-Methyl-4-ethylbenzene	1,581
1,3,5-Trimethylbenzene	1,451
1-Methyl-2-ethylbenzene	1,321
1,2,4-Trimethylbenzene	2,183
tert-Butylbenzene	2,185
Isobutylbenzene	4,197
sec-Butylbenzene	4,011
1-Methyl-3-isopropylbenzene	3,284
1-Methyl-2-isopropylbenzene	3,003
1-Methyl-3-*n*-propylbenzene	3,772
1-Methyl-4-*n*-propylbenzene	3,870
n-Butylbenzene	3,872
1,4-Dimethyl-2-ethylbenzene	3,628
1,2-Dimethyl-4-ethylbenzene	3,984
1,3-Dimethyl-2-ethylbenzene	4,345
2-Methylbutylbenzene	9,099
tert-1-Butyl-3,5-methylbenzene	6,059
n-Pentylbenzene	8,195
tert-1-Butyl-3,5-dimethylbenzene	18,260
1,3,5-Triethylbenzene	18,517
1,2,4-Triethylbenzene	16,253

for neutral organics, but the change is expected to be noticeable only if the concentration of salt exceeds 1%. Also, the presence of water-miscible polar organic solvents would result in changing properties of the matrix by reducing its polarity. In addition, swelling of the polymer with the solvent might occur for a polar coating, resulting in change of K_{fs}. However, the change is not expected to be substantial when the concentration of the solvent is below 1%.[11] When samples contain more salt and dissolved organics, but are well defined so a pure matrix can be prepared, external calibration may still be appropriate. Otherwise, a standard addition of isotopically labeled analytes should be used to compensate for variations in matrix composition.

A major analytical challenge is always associated with the analysis of samples containing solids, such as sludge. Several approaches can be implemented with SPME. Sometimes modification of the extraction conditions, such as temperature, pH, salt and other additives, facilitate displacement of analytes into the aqueous phase or headspace, resulting in similar distribution constants as those obtained for pure water. Table 5.6 illustrates the quantitation of spiked sludge under salt and low pH conditions. Salt, in this case, appears to work well as a displacing reagent for the majority of analytes. In many cases, direct extraction is not possible because of a very dirty matrix, or pH condi-

Table 5.6 Analyte Recovery from Sewage Sludge*

Compound	% Recovery[a]	% Recovery Acid + Salt
Phenol	74.2	92[b]
2-Chlorophenol	128	92[b]
o-Cresol	118	95.9
m-Cresol	95.8	97.8
p-Cresol	95.8	97.8
2,4-Dimethylphenol	104	96.5
2,4-Dichlorophenol	105	78.8
2,6-Dichlorophenol	21.3	83.4
4-Chloro-3-methylphenol	91.5	85
2,3,5-Trichlorophenol	56.1	71.3
2,4,6-Trichlorophenol	21.5	66.1
2,4,5-Trichlorophenol	64.5	66.1
2,3,4-Trichlorophenol	66.8	71.9
2,4-Dinitrophenol	2.7	111
4-Nitrophenol	38.4	118
Tetrachlorophenol Isomers	16.7	61.4
2-Methyl-4,6-dinitrophenol	0	83.1
Pentachlorophenol	8	32.2

[a] 100% recovery represents the amount recovered from laboratory water sample spiked at the same level as the sewage sample.
[b] Coeluted on GC column.
* *Source:* Ref. 21.

tions, which may damage the fiber. In such situations, the headspace mode is suitable for many applications, as described briefly in the balance of this chapter. Even semivolatiles can be analyzed by this method as long as the extraction temperature is sufficiently high and good agitation conditions are provided. Application of SPME with membrane protection can assist in the extraction of non volatile species in the presence of high molecular weight interferences, which are able to passivate the coating, such as proteins or humic material.

5.3 Solid Matrix

Accurate quantitation of target analytes in solids represents a very significant challenge to the analytical community. Although SPME cannot be used directly to extract analytes from solids, several approaches can be taken to facilitate simple sample preparation. For volatiles, the typical approach is to perform headspace analysis. To quantitatively release analytes from the matrix, the temperature needs to be increased. This facilitates higher extraction amounts and faster kinetics of the process. Loss of sensitivity associated with a decrease in the distribution constant can be compensated by cooling the

Table 5.7 Recoveries (%) of BTEX in a Clay Matrix Unver Different Extraction Conditions*

	Methods				
Compounds	I[a]	II[b]	III[c]	IV[d]	V[e]
Benzene	0.01	0.08	0.11	27	80
Toluene	0.01	0.06	0.14	11	86
Ethylbenzene	0.01	0.06	0.21	6	91
m,p-Xylene	0.01	0.07	0.22	11	93
o-Xylene	0.01	0.08	0.24	7	98

[a] Room temperature headspace sampling with a normal SPME device (without cooling).
[b] 50°C headspace sampling with a normal SPME device (without cooling).
[c] Adding 15% water to clay matrix and 50°C headspace sampling with a normal SPME device (without cooling).
[d] 250°C headspace sampling with an internally cooled SPME device.
[e] Adding 5% water to clay matrix and 170°C headspace sampling with an internally cooled SPME device. 2 min. extraction in all cases.

* *Source:* Ref. 12.

fiber. Table 5.7 illustrates extraction data obtained for BTEX spiked into a clay matrix using several extraction conditions.[12] When extraction is conducted at room temperature, the recoveries are only 0.01% or less, indicating the small amount of analyte present in the headspace. Increasing the temperature to 50°C increases the extracted amount by about an order of magnitude; however, further increase of temperature results in the reduction of recoveries, since the K_{fs} decreases.[12] Addition of water helps to displace the analytes from the surface of clay and results in further improvement by a factor of 2. The situation changes dramatically when high extraction temperature is combined with a cold coating (see Figure 2.10). In this system, the analytes are removed to the headspace and they are concentrated onto the cold fiber. The recoveries are now in tens of percent. Further increase of the recoveries, corresponding to exhaustive values, can be accomplished by adding a displacing reagent, such as water, to facilitate rapid transfer of the analytes to the headspace. The above preliminary data illustrate the effect of parameters which can be adjusted to optimize the SPME of volatiles present in solids. Table 5.8 illustrates the comparison between two exhaustive extraction methods, purge-and-trap and internally cooled SPME, for the analysis of a real clay sample. The results are similar considering the preliminary nature of this experiment.

The other successful indirect SPME approach to analysis of solids involves use of water or a polar organic solvent, such as methanol. Pure water or solvent is added to remove analytes from the matrix first, followed by solid phase microextraction. When a polar solvent is used, the extract is spiked into pure water. Low temperature water extraction followed by SPME is found to be a very useful approach for polar compounds, such as herbicides.[13] Application of methanol with water spiking, on the other hand, has been found to

Table 5.8 Concentrations (ng/g) of Native BTEX in a Real-World Clay Sample, Analyzed by Both the Purge-and-Trap GC/MSD and the Internally Cooled SPME GC/Ion Trap MS

Compounds	P&T/GC/MSD[a]	IC-SPME/GC/ITMS[b]
Benzene	—	—
Toluene	—	4.5
Ethylbenzene	7.4	10.6
m,p-Xylene	35.5	22.8
o-Xylene	4.8	7.1

[a] Analysis conducted by the Wastewater Technology Center, Burlington, Ontario.
[b] Extracted by internally cooled SPME with sampling temperature at 170°C, analyzed by a GC ion trap mass spectrometer.

be useful for analysis of volatile hydrocarbons. Table 5.9 summarizes the quantitation data obtained for the extraction of native BTEX from soils and GC/ion trap MS determination. The SPME results are compared with the standard purge-and-trap approach. In the majority of cases, the agreement between both methods is very good.[14] An interesting modification to the above procedure involves volatizing the extract followed by fiber extraction from the gaseous phase.[15] Quantitation of analytes in the gas phase is easier, as discussed in Section 5.1.

For less volatile analytes, the methanol approach described above can give good results. In addition, hot water extraction, outlined in Chapter 2 (Figures 2.11 and 2.12), is a very suitable solvent-free alternative. Table 4.8, in the previous chapter, illustrates that even a simple static system can provide good results for native PAHs in a complex matrix. As more applications of this approach develop, critical operational parameters will be better defined. Table

Table 5.9 Results of Three Analytical Methods for BTEX Applied to Four Contaminated Soils

		Compound Concentration (μg/g)[a]				
Soil	Method	Benzene	Toluene	Ethylben.	*m*/*p*-Xyl.	*o*-Xyl.
S1	P&T/MSD	2.92	3.07	42.5	169	48.3
	SPME/ITMS	2.63	3.40	43.7	180	43.0
S3	P&T/MSD	t	0.33	24.9	60.5	2.23
	SPME/ITMS	0.32	0.47	19.8	47.0	1.69
S4	P&T/MSD	0.82	1.46	27.8	104	2.62
	SPME/ITMS	0.78	1.25	20.4	96.9	2.28
S2	P&T/MSD	w	t	0.19	0.85	0.10
	SPME/ITMS	0.02	0.06	0.17	0.78	0.09

[a] Key: w, not detected; t, detected but below the statistical instrument detection limit.

Table 5.10 Amount Extracted by SPME from Collected Water under Different Conditions After Dynamic High Temperature Water Extraction as % of Amount Extracted by SPME from Spiked Water (% RSD)[a]

	Direct SPME from Spiked Water A	No Cooling, SPME After Water Extraction B	Cooling, SPME After Water Extraction C	No Cooling, SPME Simultaneous With Water Extraction D	Cooling, SPME Simultaneous With Water Extraction E
Naphthalene	100 (4)	19 (18)	110 (8)	19 (18)	96 (7)
Anthracene	100 (5)	70 (11)	99 (10)	76 (8)	97 (5)
Fluoranthene	100 (6)	101 (6)	107 (6)	102 (6)	111 (8)
Pyrene	100 (9)	102 (6)	103 (6)	102 (10)	111 (8)
Benzo(a)pyrene	100 (4)	158 (29)	50 (8)	162 (18)	102 (7)

[a] Sample: spiked sand. All experiments were performed in triplicate. Amount spiked to sand equals amount spiked to water. A: extraction with SPME fiber was performed on spiked 30 ml water, without high temperature water extraction. B: Collection device was without cooling bath and extraction with SPME fiber after the water extraction was complete. C: Collection device was with cooling bath and extraction with SPME fiber after the water extraction was complete. D: Collection device was without cooling bath and with simultaneous extraction with SPME fiber. E: Collection device was with cooling bath and simultaneous extraction with SPME fiber. For B, C, D and E, spiking was performed onto sand placed in the extraction cell. Extraction was performed at 300°C, 300 atm, and 1.2 mL/min for 15 minutes.

5.10 summarizes preliminary data obtained for a dynamic system indicating two important factors. It is important that the fiber be immersed in the aqueous phase when it is cooled down. Extraction from the hot aqueous phase results in loss of more volatile analytes. On the other hand, when the fiber is not immersed continuously in water during collection or cooldown, loss of poorly soluble analytes occur. Both static and dynamic approaches of hot water extraction have their advantages. The static method is very simple and inexpensive since it does not use high pressure pumps. The dynamic approach, on the other hand, provides extract which is much cleaner than the original matrix. However, it might be possible to extract many semivolatile target analytes from very dirty matrices with the high temperature static system, when using the headspace mode of SPME.

5.4 Environmental

A lot of research has been done on samples of environmental origin: air, water, sludge, and soils. The majority of applications have been developed for aqueous matrices. The results obtained for priority pollutants in water are very encouraging, indicating that the performance of SPME can meet USEPA method requirements. Table 5.11 summarizes the limit of detection (LOD) results obtained for solid phase microextraction of volatiles and semivolatiles, including phenols. Data for pesticides is available in Table 5.15. In each case,

Table 5.11 Summary of Detection Limits for SPME of Analyte

Analyte	LOD (pg/mL)	EPA (pg/mL)
BTEX	1–15	30–90
Polychlorinated HC solvents	1–100	10–100
PAHs	1.2–20	40
PCBs	3	60–100
Phenols	10–800	1,500–42,000

the SPME method can be optimized to meet regulatory agency requirements. The low detection limits reflect the fact that all extracted analytes are introduced to analytical instrument for determination. Below are the most important accomplishments obtained for several groups of analyte.

Benzene, toluene, ethylbenzene and xylenes (BTEX) have been the most investigated group of organic compounds of environmental interest. In many cases, performance of SPME for various matrices has been tested using these

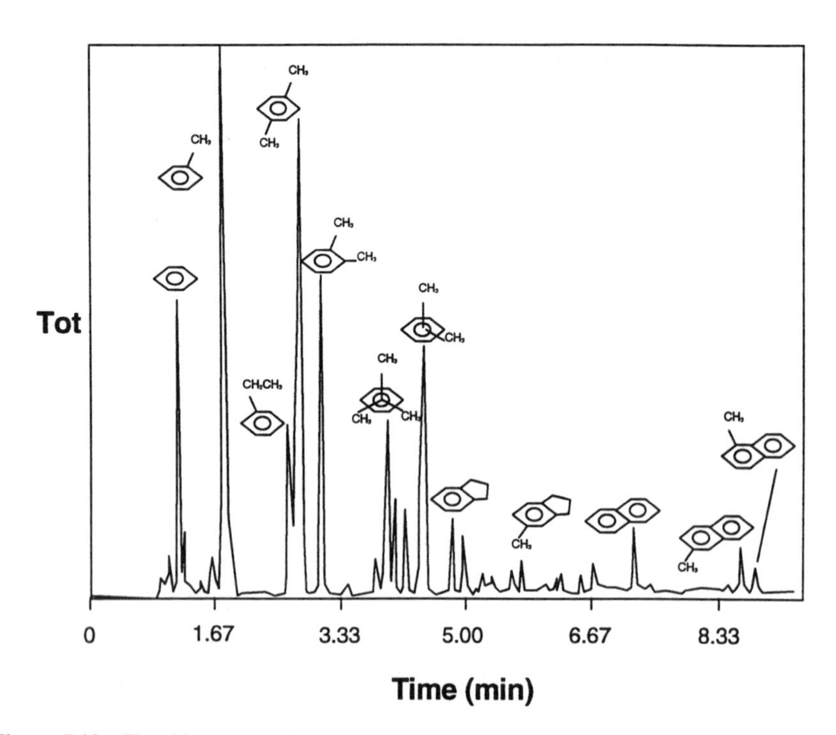

Figure 5.12 Total ion chromatogram obtained by exposing the fiber to water contaminated by gasoline components.

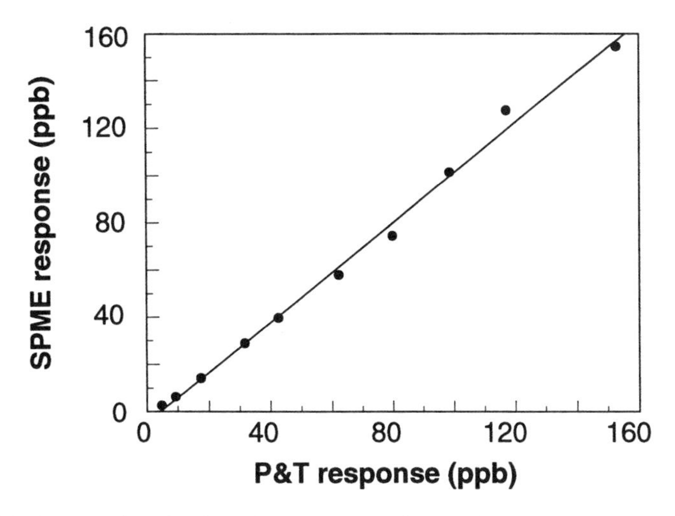

Figure 5.13 Regression plot for toluene. Each point on the graph represents a comparison between headspace SPME and the purge-and-trap technique for the same water sample.

Source: Adopted with permission from ref. 16.

analytes (see Tables 5.3, 5.7, 5.8, 5.9). As shown above in Table 5.9, the sensitivity of the SPME/PDMS method for these compounds is very high, allowing mid- to low-ppt determinations in water. This facilitated immediate application of the technique to practical environmental samples which provided a significant momentum to the early investigations. Figure 5.12 shows one of the first applications, the extraction of an aqueous matrix contaminated by gasoline components. BTEX compounds are volatile, so the headspace SPME mode can be used successfully in many applications. Careful validation

Table 5.12 Results of Two Analytical Methods for BTEX Applied to Four Nonspiked Municipal Sludge

		Compound Concentration (μg/L)[a]				
Sludge	Method	Benzene	Toluene	Ethylben.	*m*/*p*-Xyl.	*o*-Xyl.
Guelph	P&T/MSD	w	24.6	6.48	13.8	4.50
	SPME/ITMS	2.70	17.7	10.9	10.5	7.89
Calgary	P&T/MSD	2.52	50.0	36.4	231	103
	SPME/ITMS	3.18	31.8	35.1	167	123
Winnipeg	P&T/MSD	w	5.96	5.22	21.7	10.5
	SPME/ITMS	2.04	4.88	6.86	27.9	12.1
Halton	P&T/MSD	3.00	900	20.6	62.4	29.4
	SPME/ITMS	2.94	*	19.8	31.4	31.5

[a] Key: w, not detected; *, not calculated—out of range for isotope dilution.

Table 5.13 Analytical Characteristics for Analysis of Selected Polar Compounds in Water by SPME

Compound	RSD (%) (100 ppb)	LOD (ppb)
Acetone	3.9	5
Tetrahydrofuran	0.8	1
Methylethylketone	0.8	1
i-Propanol	1.9	2
t-Butanol	3.2	5
Methylisobutylketone	2.9	0.1

research has been conducted to compare performance of this method with more established alternatives. Figure 5.13 illustrates an example of a multilevel validation for toluene. The correlation line, which closely follows y = x, indicates the equivalence between the purge-and-trap technique and SPME.[16] The validation work has been extended to sludge and soils. The results are summarized in Tables 5.8, 5.9, and 5.12. Considering the complexity of the

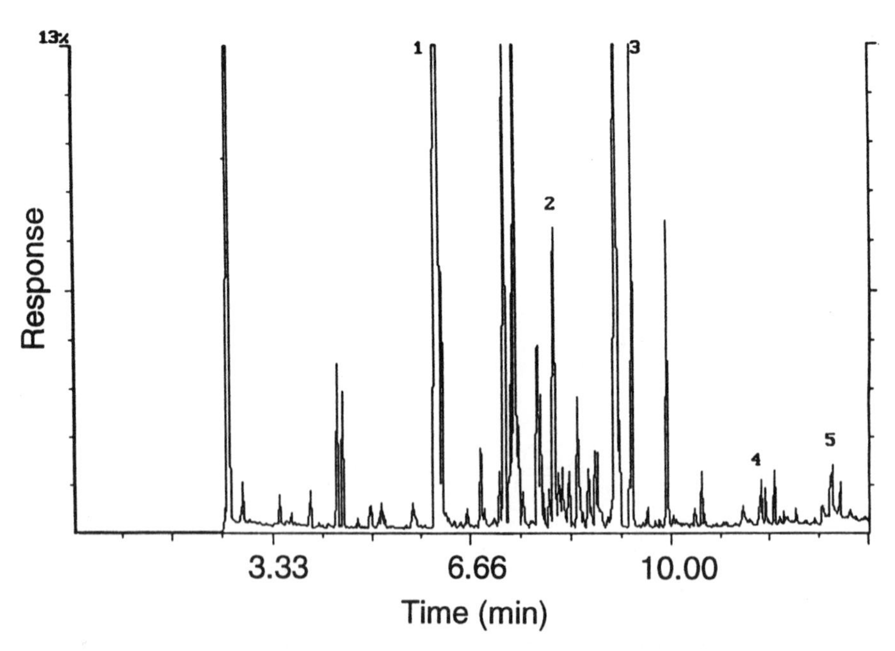

Figure 5.14 Total ion chromatogram showing organics detected and identified in a paper mill discharge wastewater sample extracted by headspace SPME at 25°C for 30 minutes: naphthalene (1), 1,1′-biphenyl (2), dibenzofuran (3), anthracene (4), dibutyl phthalate (5).

matrices, a good correlation between the Purge and Trap and headspace SPME techniques can be noted. Validation work with other volatile organic compounds, such as chlorinated hydrocarbons, produces similar conclusions.[17]

The measure of **total petroleum hydrocarbons (TPH)** is frequently used to estimate the contamination level of a sample. For air samples, this value can be calculated directly from chromatographic data without need for identification of all individual components, as emphasized in Section 5.1. TPH can then be reported, together with more specific information related to distribution and type of contaminants present in the sample, without performing an additional experiment. TPH of water samples is possible if the groups of compounds are separated and identified by a chromatographic technique combined with a quantitation method such as mass spectrometry (see the discussion in Section 5.2). For solids, this value can be obtained after the analytes are transferred into the gaseous or aqueous phase during the extraction process, as described in Section 5.3.

Polar volatile solvents like alcohols, ketones and aldehydes are difficult to quantify at trace levels in aqueous matrices. Extraction with nonpolar solvents leads to poor recoveries. More polar solvents cannot be used since they are miscible with water. SPME with a polar polymeric coating has the potential to provide a needed alternative. Table 5.13 illustrates that low ppb detection limits are possible even with FID when PDMS/DVB coating and 35% salt concentration are used.[18]

Table 5.14 The Measured Distribution Constants K Between PA Fiber and Water, the Octanol-Water Partition Coefficients (log K_{ow}), the Aqueous Solubility (mg/L), the Limit of Detection (LOD) of Heteroaromatic Compounds by SPME/GC/FID and SPME/GC/ITMS in μg/L and the % RSD

Compound	K	log K_{ow}[a]	S_w	LOD-FID μg/L	LOD-MS μg/L	% RSD ITMS
Thiophene	0.34	1.81	3,600	n.d.	1	14
1-Methylpyrrole	0.04	—	soluble	n.d.	2.5	13
Pyrrole	0.1	0.75	16,000	n.d.	10	12
2-Methylpyridine	0.02	1.06	soluble	n.d.	10	5.9
2,4-Dimethylpyridine	0.09	—	soluble	n.d.	10	14
Benzofuran	3.1	2.67	224	3	0.03	4.2
Benzothiophene	7.9	3.12	130	2	0.02	5.8
Quinoline	0.46	2	3,000	15	0.3	10
Indole	3.2	2.03	6,500	2	0.02	6.9
2-Methylquinoline	0.53	2.23	—	10	0.2	3.3
Dibenzofuran	22	4.12	10	2	0.03	10
Dibenzothiophene	27	5.45	1	2	0.02	11
Acridine	7.7	3.71	1.2	0.5	0.02	9.2
Carbazole	29	3.5	38	0.5	0.02	10
DBT-sulfone	4.9	—	—	0.5	0.04	9.1

[a] "unknown" indicated by a dash.

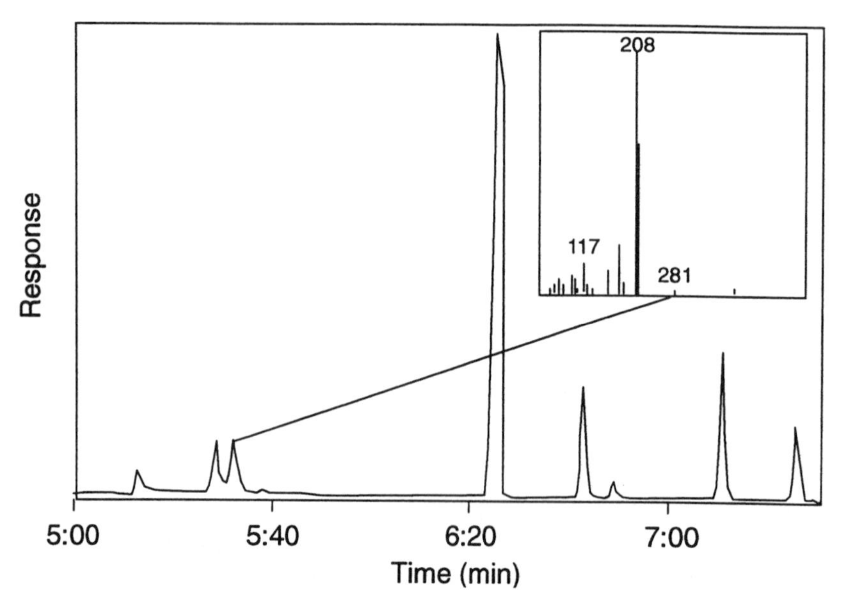

Figure 5.15 Total ion current GC/ITMS chromatogram showing the presence of methylamine in a wastewater sample after derivatization with pentafluorobenzaldehyde (PFBAY). The mass spectrum corresponds to PFBAY imine of methylamine.

Nonpolar semivolatiles, such as PAHs and PCBs are characterized by very large K_{fs} which result in very high sensitivities of SPME determination, frequently reaching low ppt levels.[19,20] Figure 5.14 illustrates major components detected and identified in a pulp mill discharge by SPME/GC/ITMS. The major advantage of SPME, compared to alternative techniques, is its field portability. Frequently, it is very difficult to obtain and preserve a representative sample since nonpolar semivolatiles have a tendency to adsorb on solid surfaces associated with the sampling vessel or on particulate matter. By sampling directly in the field, these limitations can be substantially reduced.

Polar semivolatiles in natural matrices represent a very significant analytical challenge. As discussed in Chapter 4, adjustments of pH and salt addition help to reach the required detection limits. For example, LODs obtained for **phenols**[21] with direct extraction from an acidified solution with a pH stable poly(acrylate) coating are lower than values specified by the EPA (see Table 5.14). Derivatization with acetic anhydride allows further improvement in sensitivity and chromatographic performance. The derivatization approach also enhances the extraction performance of **carboxylic acids**[22] (see Chapter 2) and **amines**, particularly for lower molecular weight species. Figure 5.15 shows the results of SPME extraction involving derivatization with pentafluorobenzylaldehyde of a wastewater sample containing methylamine.[23] Table

Table 5.15 Detection Limits for SPME Coupled with Various Detectors

Pesticide Class	Target Analyte Subgroups	ECD (ng/L)	NPD (ng/L)	MS (ng/L)	EPA Methods (ng/L)
Nitrogen-containing herbicides[a]	Thiocarbamates		20–60	0.05–.8	100–200
	Triazines		40–6000	0.4–3	100–800
	Nitroanilines		10–30	0.02–0.4	200
	Substituted uracils		200–400	0.1–1	2500–4500
	Substituted amide		800	15	500
	Acetanilide		200	0.01	700
	Diphenyl ether		300	6	N/A
	Triazole		30	0.01	N/A
Organochlorine pesticides[b]	BHCs	0.9–9[c]		0.01–0.04	10–25
	Hexachlorocyclodienes	0.06–1.6		0.02–0.6	2.5–50
	Diphenyl aliphatics	0.05–4.7		0.06–4.5	10–75
Phosphorous-containing pesticides[a]	Phosphate		500	6	100–2500
	Phosphorothiolate		130	2	N/A
	Phosphonothiate		15	8	9–200
	Pyrophosphate		16	1	6
	Phosphorodithioates		9–320	0.4–9	9–1500
	Phosphorothioates		11–280	0.7–100	4–2000

[a] Analyzed by SPME with 85 μm poly(acrylate) fiber.
[b] Analyzed by SPME with 100 μm poly(dimethylsiloxane) fiber.
[c] Hexachlorocyclohexanes (BHCs) are analyzed with 30% w/w salt added.

5.14 shows the results obtained for **very polar heteroaromatic** compounds frequently present in creosote contaminated groundwater.[24] Extraction with a PA coating, combined with sensitive ion trap mass spectrometry detection, allows low to sub-ppb detection limits for these very water-soluble compounds.

Pesticides are an analytically difficult group of compounds because of their wide range of chemical structures and properties. Some of them are classified as nonpolar (organochlorine pesticides) and others as very polar semivolatiles (herbicides). Despite this diversity, SPME has been very successfully applied for their determinations in aqueous matrices by several research groups.[25,26,27,28,29] Table 5.15 summarizes the SPME detection limits obtained for several groups of pesticides. They are lower compared to USEPA method LODs. It is even possible to develop single extraction conditions to facilitate screening of 60 target pesticides combining three classes: organochlorines, organophosphorous and nitrogen containing herbicides.[28] Figure 5.16 illustrates extraction results of spiked water containing all the analytes.

In addition, the round robin test discussed in Chapter 5 demonstrates the ruggedness of the SPME/GC pesticide method, and its ability to be rapidly implemented in different laboratories. Many pesticides are nonvolatile compounds requiring direct extraction, which results in relatively long equilibration times, frequently exceeding 60 minutes with magnetic stirring. Increasing

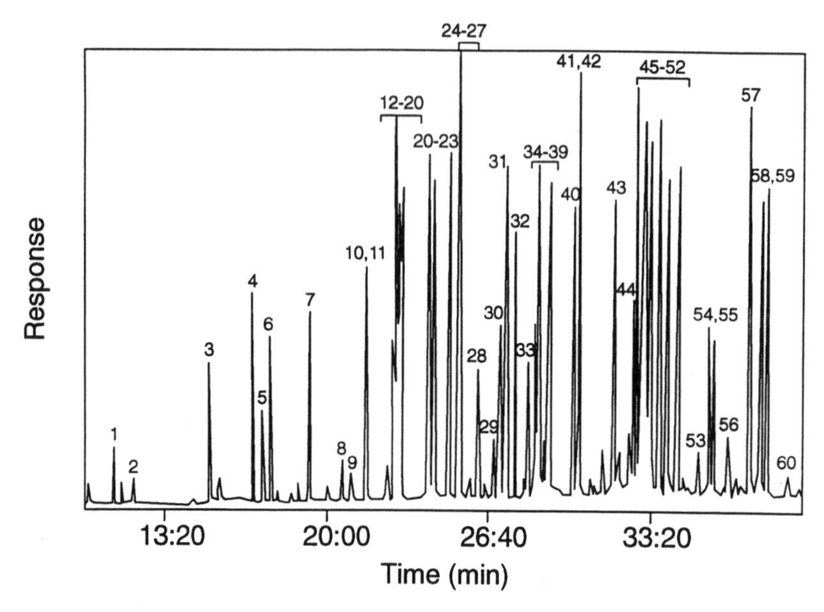

Figure 5.16 SPME/GC/ITMS total ion current chromatogram obtained after extraction of 60 pesticides from aqueous sample. Conditions: 5 min, ramped at 30°/min to 100°C, ramped at 5°/min to 275°C and ramped at 30°/min to 300°C; a 30 m × 0.25 mm i.d. SPB column. 1 - *o,o,o*-TEP, 2 - Dichlorovos, 3 - EPTC, 4 - Butylate, 5 - Vernolate, 6 - Pebulate, 7 - Molinate, 8 - Thionazin, 9 - Propachlor, 10 - Ethoprofos, 11 - Cycloate, 12 - Trifluralin, 13 - Benfluralin, 14 - Sulfotep, 15 - Phorate, 16 - α-BHC, 17 - Dimethoate, 18 - Simazine, 19 - Atrazine, 20 - β-BHC, 21 - Propazine, 22 - Lindane, 23 - Profluralin, 24 - Diazinon, 25 - δ-BHC, 26 - Disulfoton, 27 - Terbacil, 28 - Iprofenfos, 29 - Metribuzin, 30 - Methyl Parathion, 31 - Heptachlor, 32 - Fenchlorovos, 33 - Fenitrothion, 34 - Bromacil, 35 - Isoxathion, 36 - Aldrin, 37 - Metolachlor, 38 - Chloropyrifos, 39 - Ethyl Parathion, 40 - Isopropalin, 41 - Pendimethalin, 42 - Heptachlor epoxide, 43 - Endosulfan I, 44 - Prothiofos, 45 - *p,p'*-DDE, 46 - Dieldrin, 47 - Oxadiazon, 48 - Oxyfluorofen, 49 - Endrin, 50 - Endosulfan II, 51 - *p,p'*-DDD, 52 - Endrin aldehyde, 53 - Famphur, 54 - Endosulfan sulfate, 55 - *p,p'*-DDT, 56 - Hexazinone, 57 - Endrin ketone, 58 - EPN, 59 - Methoxychlor, 60 - Azinphos-methyl.

Source: Adopted with permission from ref. 27.

temperature will decrease the equilibration times, but cooling the fiber might be necessary to maintain good detection limits. To date, only GC separation methods have been used, limiting application of the SPME technique to thermally stable pesticides. Application of HPLC would allow analysis not only of less stable species such as carbamates, but also products of pesticide metabolism and degradation.

Application of sodium teraethylborate as an *in situ* derivatization reagent results in quantitation of some **metal ions and organometallic species** in aqueous samples by SPME. This approach was first used for the determination of

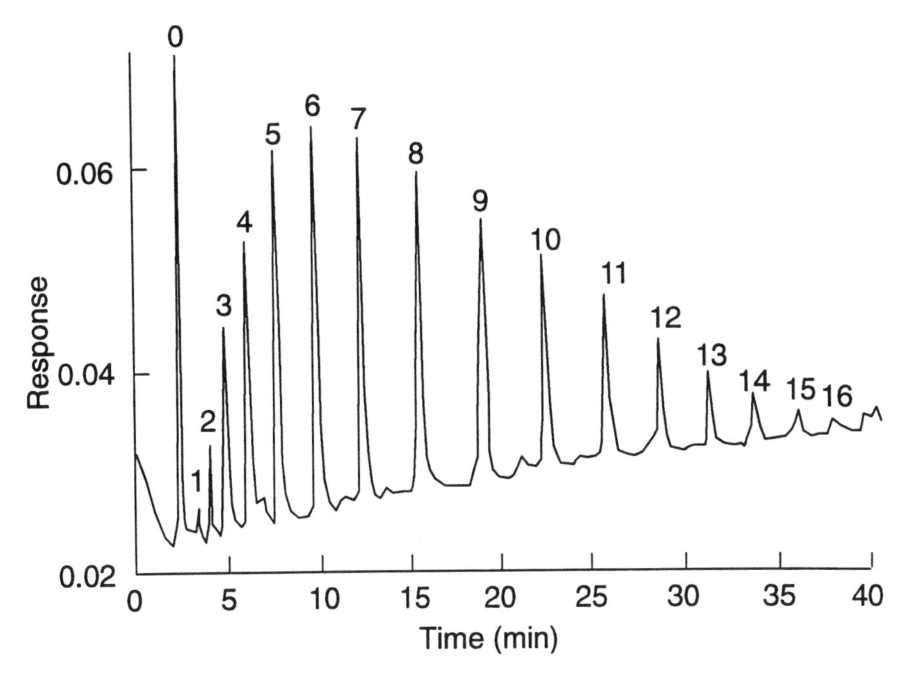

Figure 5.17 HPLC chromatogram of the extracted alkylphenol ethoxylates from Triton X-100 in water. Peak assignments correspond to number of units in the ethoxylate chain.

Source: Adopted with permission from ref. 32.

methylmercury in water and biota,[30] followed by the application of this approach to quantitation of lead ion and organolead compounds.[31] In the method, analytes are first derivatized to fully ethylated forms, before they are extracted by the PDMS fiber. These species are volatile, therefore, the headspace extraction mode of SPME can be applied, facilitating rapid analysis of very complex matrices. The reaction time, which takes about 15 minutes, typically determines the extraction time. Stirring must be very efficient during derivatization, since the reagent decomposes rapidly after contact with the sample. The affinity of neutral forms of organometallic compounds toward a PDMS coating is very high, resulting in very low detection limits, approaching sub-ppt levels for determination of tetraethyllead in water by SPME/GC/ITMS. This method can be applied to partial speciation of various forms of these metals present in a sample, first by extracting native neutral organometallic species followed by determination of the total metal content in the sample. Full speciation can be accomplished by isotopically labeling the derivatization reagent, and then differentiating between various forms based on their mass spectra.[32]

Application of HPLC to separation of extracted species by SPME opens new possibilities in the determination of nonvolatile and thermally unstable environmental pollutants. Figure 5.17 illustrates chromatograms obtained for several alkylphenol ethoxylates **non-ionic surfactants** after extraction from water[33]. Selection of the appropriate coating and optimization of the desorption conditions were discussed in Chapter 4. The detection limits obtained are in the low ppb level for individual ethoxamers with UV detection, which indicates good prospects for determination at the trace level in natural matrices.

5.5 Food and Pharmaceuticals

A unique feature of SPME for application to analysis of food and drugs is its ability, in some cases, to extract substances from products without opening the package. For example, flavors present in wine can be checked before sale or purchase by introducing a fiber into the headspace of the wine through the cork of the bottle. Since only an insignificant amount of flavor is extracted, the composition of the product does not change, and the extracting solvent does not contaminate the product since the polymeric phase is not volatile. A similar process can be applied to on-line product monitoring of each individual item to ensure the best possible quality.

Food and pharmaceutical products are frequently contaminated by volatile **organic solvents**, which are used in manufacturing and processing of these

Table 5.16 Linear Range and Limits of Detection (LOD) and Limits of Quantitation (LOQ) for the Orange Juice Flavour Volatile Compounds Using the Poly(acrylate) Coated Fiber

Target Analyte	Linear Range (ppm)	LOD (ppm)	LOQ (ppm)
Methanol	1.4×10^2—0.14	3.5×10^{-2}	1.2×10^{-1}
Ethanol	9.4×10^2—0.94	3.9×10^{-1}	9.0×10^{-1}
Ethyl acetate	2.3–0.023	3.9×10^{-3}	1.3×10^{-2}
2-Methyl-1-propanol	26–0.027	4.1×10^{-3}	1.4×10^{-2}
Methyl butyrate	1.2–0.0012	4.7×10^{-4}	1.0×10^{-3}
Ethyl butyrate	0.55–0.00055	8.1×10^{-5}	2.7×10^{-4}
cis-3-Hexen-1-ol	0.79–0.00079	1.8×10^{-4}	6.2×10^{-4}
Hexyl alcohol	0.75–0.00075	9.0×10^{-5}	3.0×10^{-4}
α-pinene	0.084–0.00084	2.7×10^{-5}	9.0×10^{-5}
β-Myrcene	0.013–0.00013	6.8×10^{-6}	2.7×10^{-5}
Ethyl hexanoate	0.015–00015	3.8×10^{-6}	1.3×10^{-5}
Octanal	0.014–0.00014	3.8×10^{-6}	1.3×10^{-5}
Limonene	0.79–0.0079	4.7×10^{-6}	1.6×10^{-5}
γ-Terpinene	0.81–0.0081	3.8×10^{-6}	1.3×10^{-5}
Linalool	0.77–0.00077	4.1×10^{-6}	1.4×10^{-5}
α-Terpineol	0.71–0.0071	3.6×10^{-7}	1.2×10^{-3}
Decanal	0.7–0.0070	1.3×10^{-6}	4.3×10^{-6}

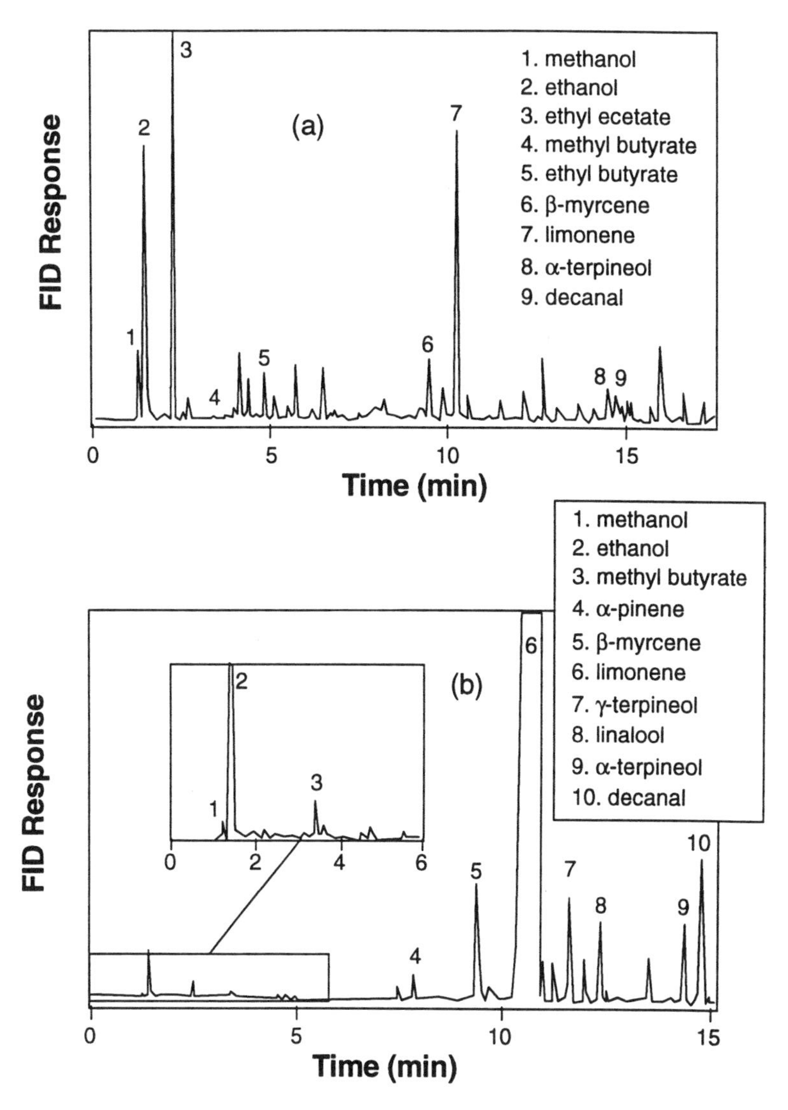

Figure 5.18 SPME/GC/FID chromatogram of a headspace of (a) Welch's Grape Juice and (b) a concentrated grapefruit juice. Fiber coating: poly(acrylate).

Source: Adopted with permission from ref. 37.

goods. Considering the frequent complexity of their matrices, the analytical procedures can be quite involved. Static headspace is frequently used for this purpose, but SPME has been found to be a good alternative for analysis of both contaminated pharmaceuticals[34] and foods[35] because of its superior sensitivity especially for less volatile compounds. Samples can be analyzed

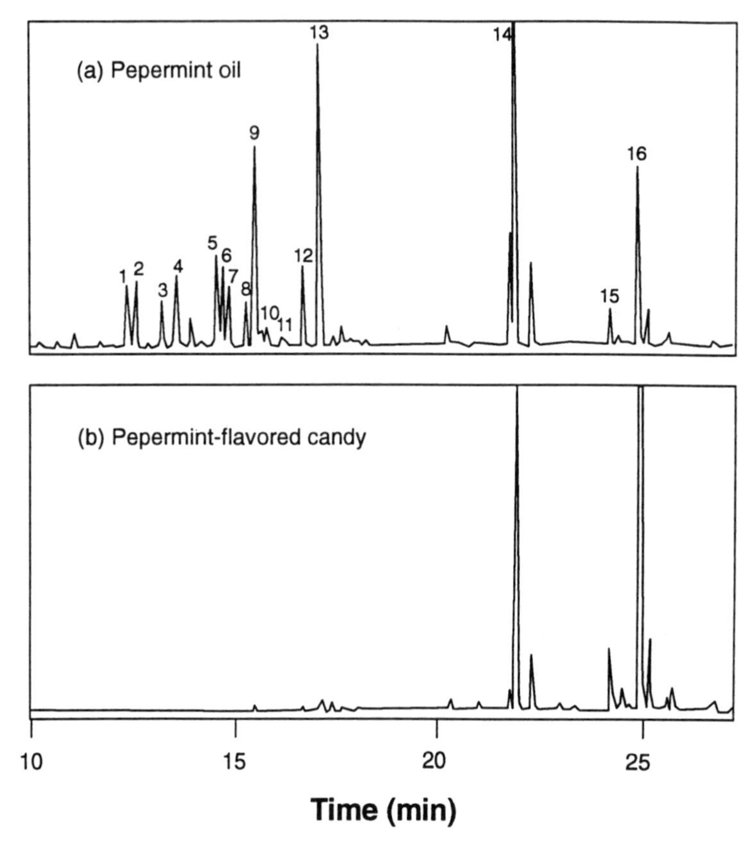

Figure 5.19 Headspace SPME extraction and chiral GC analysis of (a) peppermint oil and (b) peppermint-flavored candy using a β-cyclodextrin column. Oven temperature program: 40°C for 2 minutes, then 40–220°C at 4°C/min. Peaks: 1, (−)α-pinene; 2, (+)α-pinene; 3, β-myrcene; 4, (+)-sabinene; 5, (−)-camphene; 6, (−)β-pinene; 7, (+)-β-pinene; 8, α-terpinene; 9, 3-carene; 10, (−)-limonene; 11, (+)-limonene; 12, (±)β-phellandrene; 13, γ-terpinene; 14, cineole; 15, menthone; 16, (+)-menthol.

Source: Adopted with permission from ref. 38.

by dissolving the product in water prior to determination, or the analytes can be extracted directly from the headspace above the product. Typically, standard addition is used for quantitation. The detection limits are similar to those obtained for environmental matrices.

Flavors are another very interesting topic of investigation. One strength of SPME is its small size which enables it to extract compounds from small objects such as a single pill or small botanical. SPME has been used on a wide range of food products, spices, oils, and beverages,[36] and even black and white truffle aroma.[37] Table 5.16 summarizes the detection limits of common flavor components obtained by a PA fiber.[38] Figure 5.18 illustrates examples of

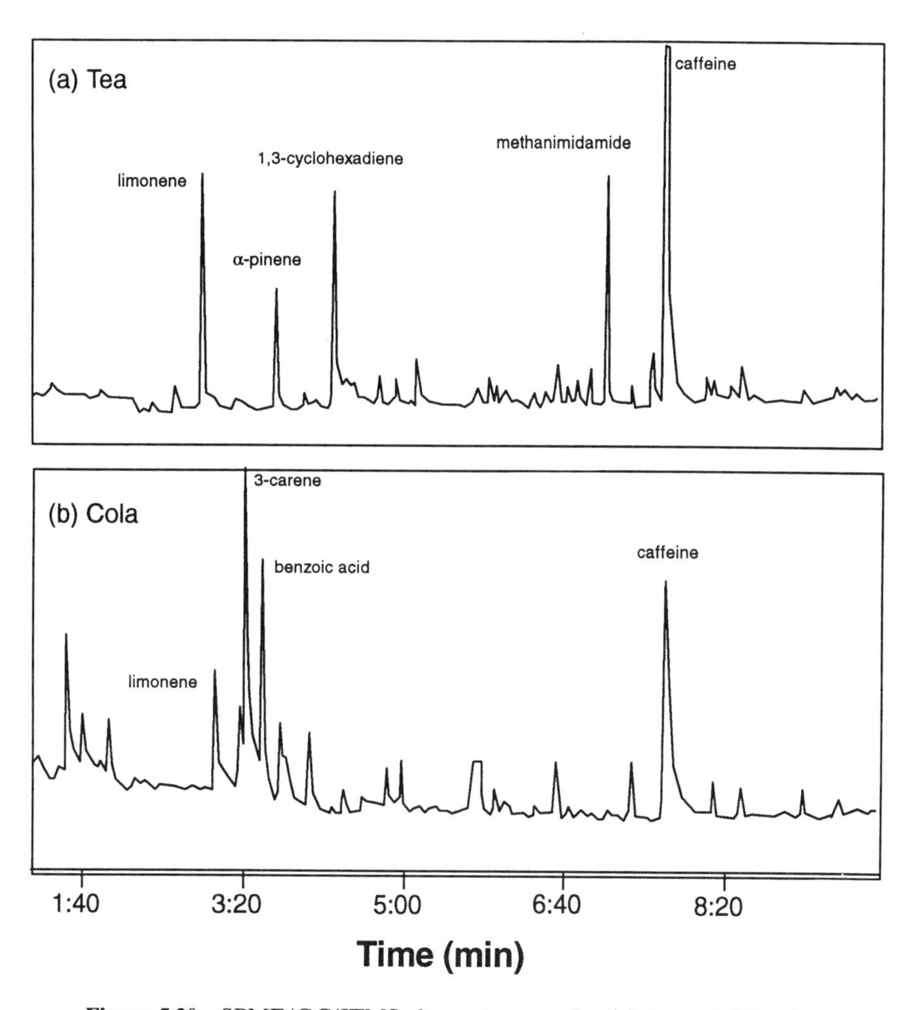

Figure 5.20 SPME/GC/ITMS chromatograms for (a) tea and (b) cola.

chromatograms obtained for two fruit juices. Initial investigations indicate that good quantitation of volatile analytes in the matrix can be obtained after the solid components of the sample matrix are separated from the aqueous phase by centrifuging. Combining SPME extraction with GC separation, using the chiral stationary phase, provides information not only about the composition of the products, but also their natural purity.[39] Figure 5.19 shows the separation of peppermint oil (a) and peppermint-flavored candy (b) using a β-cyclodextrin column.

Another interesting application of SPME to analysis of foods, is to quantify **caffeine** in soft drinks.[40] Figure 5.20 shows the results obtained for tea and cola. The most convenient quantitation method for this determination is based

Table 5.17 Herbicides Found in a Wine Sample

Herbicide	Concentration (ng/mL)[a]
Butylate	0.7
Pebulate	3.7
Trifluralin	2.2
Benfluralin	2
Profluralin	3.1
Isopropalin	0.8
Pendimethalin	0.9
Oxadiazon	0.4
Oxyfluorofen	1

[a] Triplicate samples quantified by standard addition at three concentrations.

on isotopically labelled standards which are readily available. Figure 5.20 indicates that in addition to caffeine, flavour components are also detected in the same chromatographic run, suggesting that SPME could be a very versatile quality control tool in the field of beverage production.

The number of research reports on semivolatiles in foods is more limited, but the scope of potential applications is as broad as in the environmental area. For example, Table 5.17 shows **herbicides** detected a wine sample.[13]

5.6 Clinical and Forensic

The major advantage of SPME for clinical and forensic applications is its portability, which might eventually be translated into an on-site sampling/ sample preparation followed by immediate analysis on portable instruments or transport to the laboratory. This approach would allow better monitoring of patients' conditions during treatment or therapy, and better preservation of crime scenes, since the objects would not need to be removed to perform analysis.

Monitoring of human health can be achieved by noninvasive SPME methods. For example, breath contains the headspace of blood. Therefore, characterization of volatile components present inside the human body can be accomplished by adapting a SPME device to breath analysis, as illustrated in Figure 2.3a (Chapter 2). Figure 5.21 illustrates the results of an initial investigation comparing the response of different coatings at various humidities to three different components of human breath: **ethanol, acetone, and isoprene**.[41] The PDMS/DVB coating extracts the largest amount of analytes from a breath sample, but also it is very much effected by the humidity level. On the other hand, pure PDMS is the least sensitive to the polar analytes, but at the same

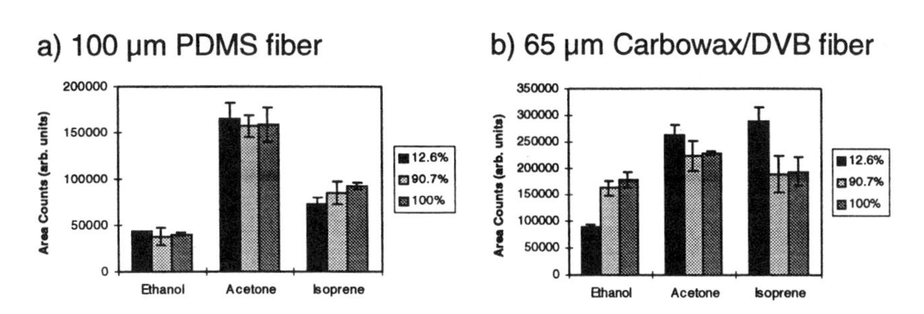

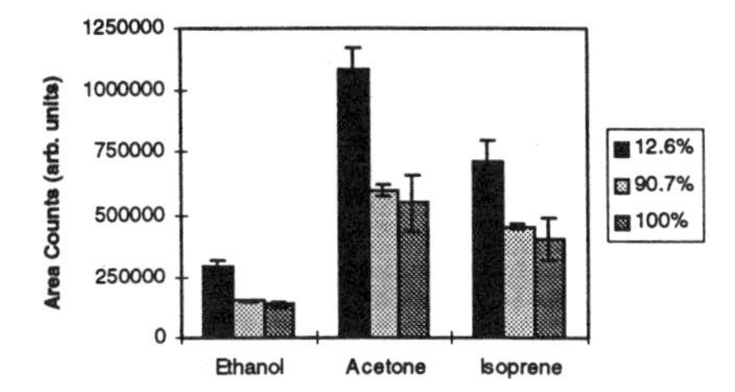

Figure 5.21 Effect of coating type and relative humidity on extraction amount from gaseous samples.

time does not exhibit substantial change in sensitivity as the humidity level varies. Figure 5.22 illustrates an example of breath analysis. The detection limits are in the low nmol/L range, which provides sufficient sensitivity for the practical applications of this approach to monitoring alcohol levels in blood and acetone concentrations for diabetics and patients on diets. SPME can also be used for rapid analysis of body fluids. Figure 5.23 illustrates the detection of several aromatic amines in the milk of a woman who is a smoker.

Analysis for **alcohol** and **drugs** in body fluids is frequently performed in clinical and forensic laboratories. SPME is suitable for monitoring alcohol levels in both urine and blood. Figure 5.24 shows the linear regression for ethanol determination in blood and urine by headspace SPME versus the static headspace technique. The results indicate that both techniques give equivalent results for this application. Drugs in body fluids are more difficult to analyze, since they are semivolatile substances. However, a number of published preliminary results indicate that headspace SPME analysis at an elevated temperature, or even direct extraction after dialysis of urine or blood serum is possible. Most of the work to date was done with **amphetamines**,[42,43] but analyses of other drugs has been also reported such as **valproic acid**,[44] **cocaine**, **tricyclic antidepressants**,[45] **nicotine**, and **local anesthetics**. For a full listing of references, refer to Appendix B.

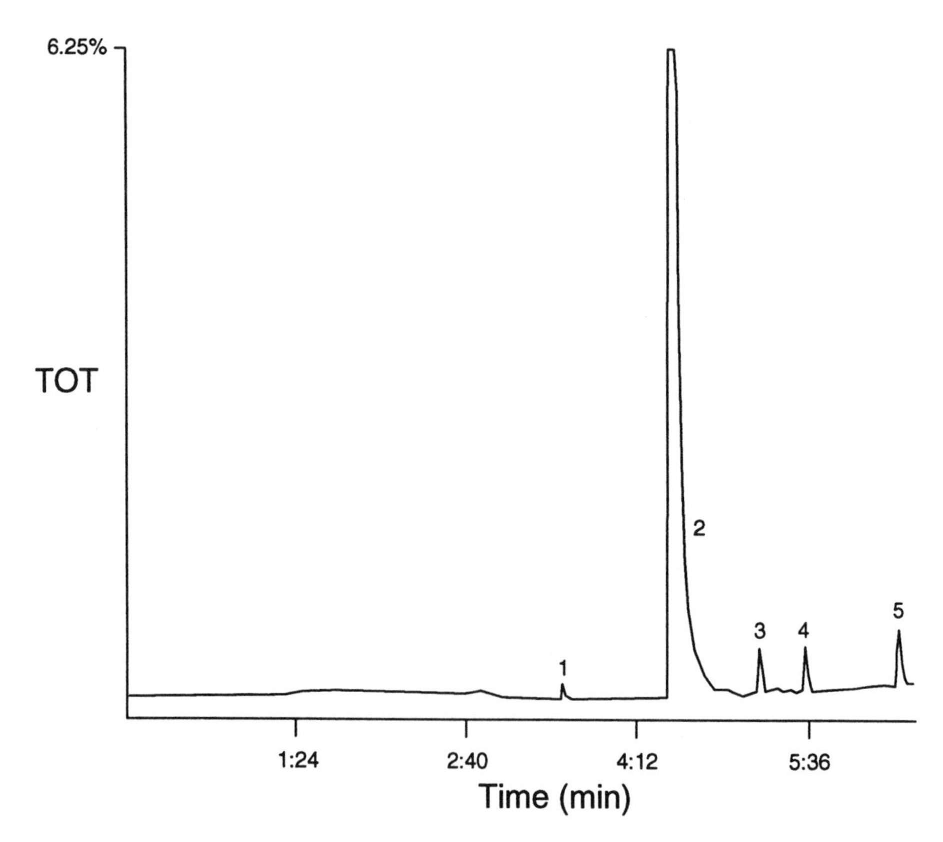

Figure 5.22 SPME/GC/ITMS analysis of a breath sample. Breath sample was taken after alcohol consumption, Fiber coating: Carbowax/DVB fiber. Peak assignment: acealdehyde (1), ethanol (2), acetone (3), isoprene (4) and carbon disulfide (5).

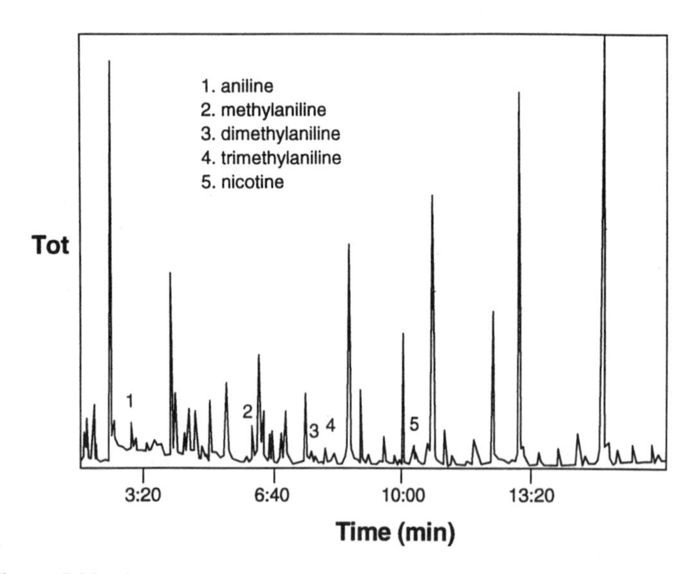

Figure 5.23 SPME/GC/ITMS analysis of human milk from a smoker.

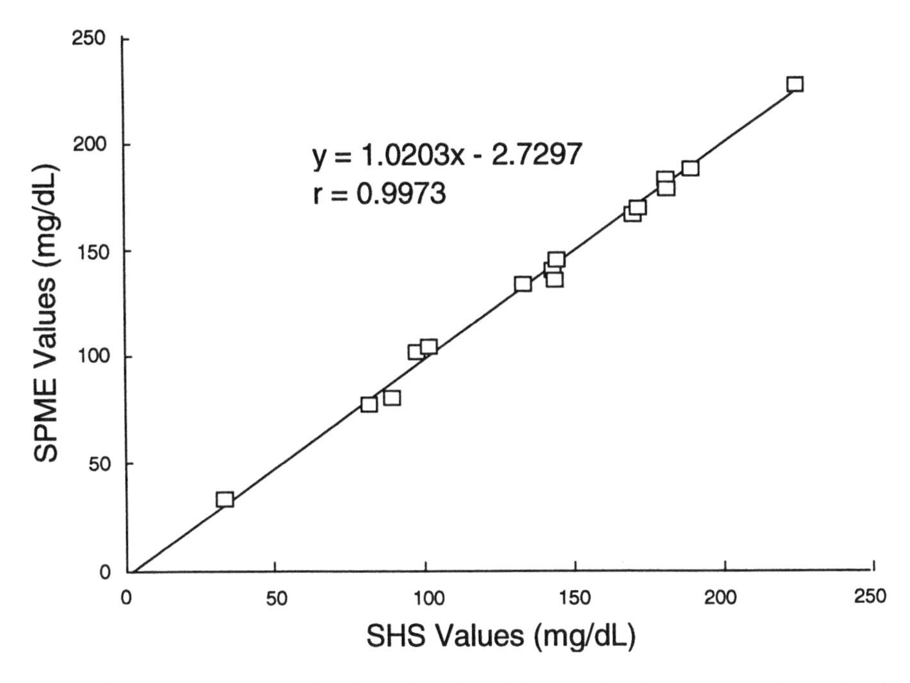

Figure 5.24 Linear regression for ethanol in blood and urine samples determined with headspace SPME versus static headspace technique. The 15 samples included 12 blood and 2 urine specimens and an aqueous control.

Source: Z. Penton, Varian.

Field analysis by SPME, facilitates the careful investigation of a crime scene. Headspace SPME has been applied to the investigation of suspected **arson** cases.[46,47] Figure 5.25 shows the selected ion chromatograms corresponding to a sample from a real arson case and a gasoline standard, indicating the possibility that an accelerant was used in the fire of suspicious origin.

5.7 Future Analytical Applications

The range of SPME applications which can be developed is limited by the availability of appropriate instrumentation and coatings. A number of the devices discussed in Chapter 2 are not commercially available yet. For example, Figure 5.26 demonstrates that 3 minute analysis time of water samples for **volatiles** is possible when using SPME as an extraction and injection technique with a flash desorption injector. Automation of this process would result in a **ten fold increase** of laboratory throughput compared to the currently used purge-and-trap systems.

A major advance over the current practice would be to move sample preparation and, possibly, analysis to a site where the sample is typically

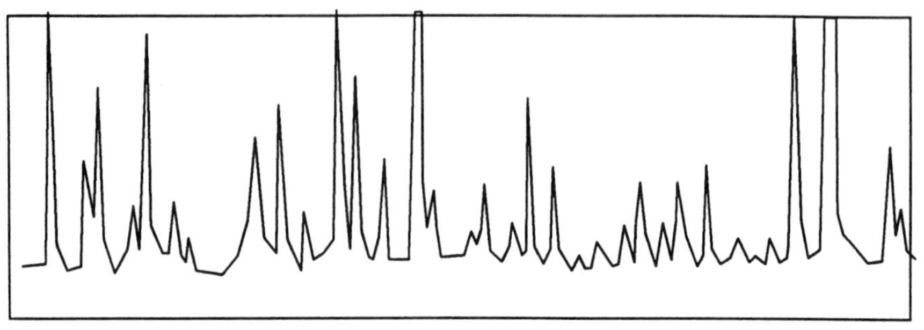

(b) SIC of a pure gasoline sample (m/z=57)

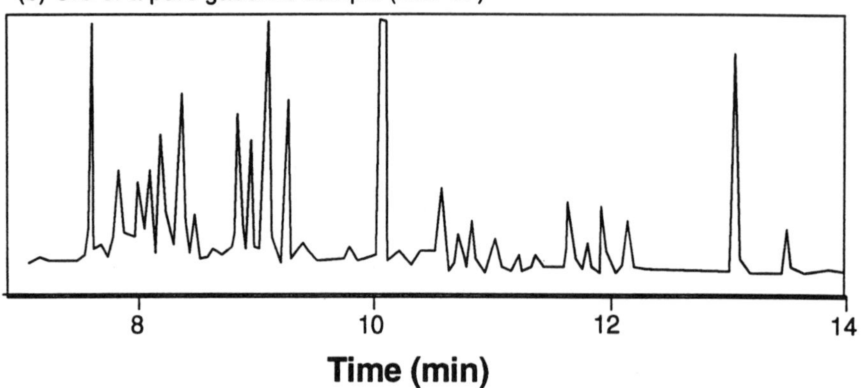

Figure 5.25 Selected ion trace for m/z = 57 for the sample obtained by using headspace SPME and GC/ITMS: (a) real fire debris sample; (b) pure gasoline.

collected. SPME field devices would reduce cost and possibility of errors associated with handling and storing samples. In addition, **field analysis** would facilitate faster and better characterization of the problem, since analytical information would be available immediately for evaluation and decision. On-site analysis may someday include **industrial hygiene** monitoring devices to facilitate the protection of workers. SPME devices could be distributed in predetermined places within the monitoring parameter, then collected and transferred periodically to a central location for analysis and data evaluation. Another future application of SPME could involve the evaluation of biotoxicity of various environments. The SPME fiber coating could be designed to

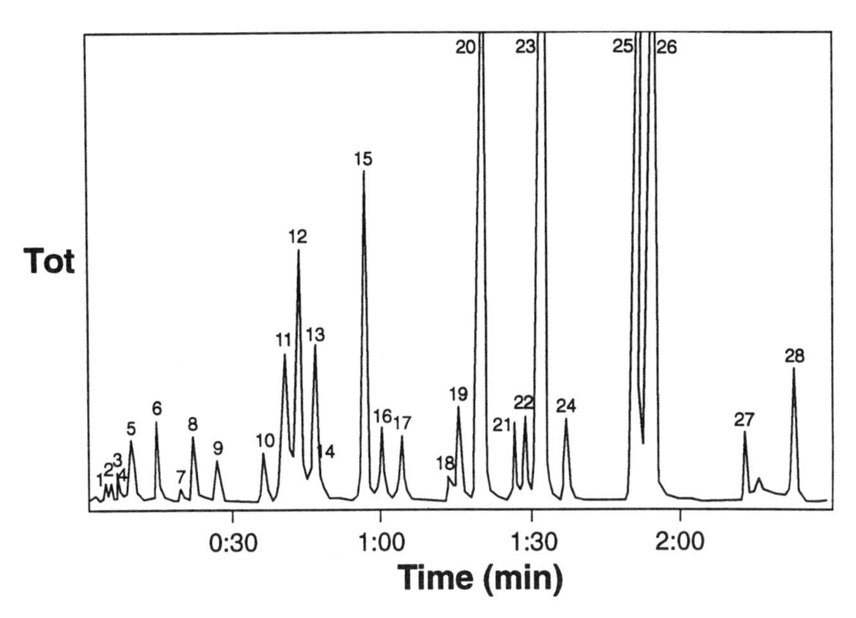

Figure 5.26 Separation of purgeables A, B and C on a VOCOL column. Conditions: 0°C-30°/min-70°; 2.1 atm, dedicated injector, capacitor voltage 24 V, MS detector, mass range 45–250. 1 - chloromethane; 2 - vinyl chloride; 3 - bromomethane; 4 - chloroethane; 5 - trichlorofluoromethane; 6 - 1,1-dichloroethene; 7 - dichloromethane; 8 - 1,2-dichloroethene; 9 - 1,1-dichloroethane; 10 - trichloromethane; 11 - 1,1,1-trichloroethane; 12 - tetrachloromethane; 13 - benzene; 14 - 1,2-dichloroethane; 15 - trichloroethene; 16 - 1,2-dichloropropane; 17 - bromodichloromethane; 18 - 2-chloroethyl vinyl ether; 19 - cis-1,3-dichloropropene; 20 - toluene; 21 - trans-1,3-dichloropropene; 22 - 1,1,2-trichloroethane; 23 - tetrachloroethylene; 24 - dibromochloromethane; 25 - chlorobenzene; 26 - ethylbenzene; 27 - tribromomethane; 28 - 1,1,2,2-tetrachloroethane.

simulate the intake of pollutants by living organisms. After desorption and analysis, the total biotoxicity could be calculated by considering the toxicity of each extracted component.

As new coatings become available, new applications can be developed. For example, ion exchange materials and crown ether ligands have been used to selectively extract **inorganic ions** from an aqueous matrix, allowing not only quantitation, but also **speciation**.[48,49] A polyacrylic acid coating has been developed for the analysis of basic **proteins**.[50] Availability of bioaffinity coatings would facilitate specific extraction of very difficult analytes from very complex matrices such as **biological samples**, or even single cells, if the fibers are designed to have small dimensions.

SPME can be applied to the majority of sample preparation tasks as long as appropriate coatings and instrumentation are available. Judgment about its advantages to perform a given task belongs to the analytical chemist, who will evaluate performance of several alternatives before making a decision to choose the most suitable. It is hoped that this book will assist the reader to make more educated decisions about SPME.

5.8 Physicochemical Applications

Solid phase microextraction can be applied not only for extraction purposes, but also to perform measurements which better characterize the extraction system. The measurements can include studying properties of the fiber coating and investigation of multiphase equilibria in the matrix. These applications are discussed in detail below since they can assist the SPME optimization process. In addition, since a detailed mathematical treatment of the solid phase microextraction process is available (Chapter 3), SPME can be applied to study parameters present in these equations, such as diffusion coefficients in both coating and extracted phases, as well as various distribution constants.

5.8.1 Properties of Fiber Coatings

Distribution constants and equilibration times can provide information about fiber coating materials. Considering relatively simple procedures associated with preparing such coatings, such investigations are very useful to provide knowledge about how a polymeric phase interacts with different types of compounds. Such studies contribute to a basic understanding of solvation mechanisms. This information would aid researchers to select and synthesize the right phase for successful extraction, or separation, of complex mixtures in a time-efficient manner. In addition, accessibility of the coating for parallel investigations using various spectroscopic techniques can expand these investigations to study properties of commercial materials such as paints and other films.

For understanding solute-solvent interactions at molecular level and the thermodynamic processes involved in forming the solution, the study of infinite dilution activity coefficients of probe solutes in a polymer phase is a fundamental approach. The measurements of these coefficients of solutes in a sorbing phase are usually performed by a static gravimetric method, or by a dynamic GC method. Both these methods suffer from many disadvantages, including the large quantity of the phase required in the static method, or the need to prepare a chromatographic column coated with the phase under examination in the GC method. Nowadays, the latter method is normally used for the determination of infinite dilution activity coefficients.[51]

SPME can be another tool for studying the properties of coatings. It has some advantages of both the static and the GC methods, while minimizing their drawbacks. The phases of interest can be coated on fibers made of suitable materials (fused silica, stainless steel, etc.). The process of making fibers is much easier than that of columns. It requires very little material. The SPME device with a selected fiber coating can be used to extract a group of probe compounds, which are then separated on a standard commercially available column and quantified by a GC/MS. No extrapolation is necessary, as very low analyte amounts can be accurately determined.

SPME measurements of infinite dilution activity coefficients of probe solutes in a selected stationary phase consist of two steps. The first step is to measure the infinite dilution partition coefficients of the probe solutes between the coating and the gas phase. Then, from these partition coefficients, the corresponding activity coefficients are deduced.

For the sake of theoretical treatment, in the following discussion we assume that liquid polymeric phases are used as coatings. For other types of coatings, the general conclusions are still true except that some parameters may have different meanings. Considering a solute dissolved in a liquid stationary phase to form an infinitely dilute solution, from Henry's law we can express its equilibration vapor pressure as

$$p_1 = \gamma_p^\infty x_1 p_1^0 \tag{5.5}$$

where p_1 and p_1^0 are the solute partial and saturated vapor pressures, x_1 is the solute mole fraction in the fibre coating, and γ_p^∞ is the activity coefficient of the solute in this infinitely dilute solution. Assuming an ideal gas behaviour for the solute vapor, we have

$$\frac{n_2}{V_g} = \frac{p_1}{RT} \tag{5.6}$$

where n_2 is the molar number of the solute vapor; V_g is the volume of the gas phase; R is the gas constant; T is the temperature in Kelvin. Let us further assume that the fibre liquid coating has a volume of V_f and molar number of n_L, and that the molar number of the solute in the infinite-dilution solution is n_1; then, $x_1 = n_1/(n_1 + n_L) = n_1/n_L$. The partition coefficient of the solute, K_{fg}, can be expressed as

$$K_{fg} = \frac{n_1}{V_L}\frac{V_g}{n_2} = \frac{n_1 RT}{\gamma_p^\infty x_1 p_1^0 V_L} = \frac{n_L RT}{\gamma_p^\infty p_1^0 V_L} = \frac{\rho_L RT}{\gamma_p^\infty p_1^0 M_L} \tag{5.7}$$

where ρ_L and M_L are the density and the molecular weight of the liquid polymeric phase of the coating, respectively. Rearranging eq. 5.7 to isolate the activity coefficient gives

$$\gamma_p^\infty = \frac{\rho_L RT}{K_{fg} p_1^0 M_L} \tag{5.8}$$

The above equation was obtained by assuming ideal gas behaviour for the solute vapor. This assumption is obviously an approximation. A correction must be made by replacing p_1^0 with fugacity f_1^0

$$f_1^0 = p_1^0 \exp\left(\frac{B_{11}p_1^0}{RT}\right) \tag{5.9}$$

where B_{11} is the second virial coefficient of the solute. A further correction must be made in the second virial coefficient (B_{11}) to account for the fact the solute vapor is mixed with a carrier gas or air during a measurement. After this correction, we have

$$\ln \gamma_1^\infty = \ln \gamma_p^\infty - \frac{p_1^0(B_{11} - \nu_1^0)}{RT} = \ln \frac{\rho_L RT}{K_{fg}p_1^0 M_L} - \frac{p_1^0(B_{11} - \nu_1^0)}{RT} \tag{5.10}$$

where γ_1^∞ is the infinite dilution activity coefficients of the solute, and ν_1^0 is the bulk solute molar volume that is used here to correct the second virial coefficient. If the partition coefficient of the solute, K_{fg}, is measured, the infinite dilution activity coefficient, γ_1^∞, can be calculated from eq. 5.10.

The activity coefficient defined in eq. 5.5 is based on the molar fraction. When a stationary phase is a polymer, its molecular weight M_L is not well defined. Molar fraction is impossible to obtain in that case. To avoid this problem, we use weight fraction instead, and rewrite eq. 5.5 as

$$p_1 = {}^w\gamma_p^\infty x_w p_1^0 \tag{5.11}$$

where ${}^w\gamma_p^\infty$ is the weight fraction infinite dilution activity coefficient, and x_w is the weight fraction of the solute, $x_w = w_1/(w_1 + w_L)$. In an infinitely dilute solution, $w_1 \ll w_L$, so x_w can be expressed as

$$x_w = \frac{w_1}{w_L} = \frac{n_1 M_1}{n_L M_L} = x_1 \frac{M_1}{M_L} \tag{5.12}$$

where w_1 and w_L are the weight of the solute and stationary phase; M_1 is the molecular weight of the solute. Replacing eq. 5.11 with eq. 5.12 and comparing eqs. 5.5 and 5.11, we have

$$\gamma_p^\infty = {}^w\gamma_p^\infty \frac{M_1}{M_L} \tag{5.13}$$

Rewrite eq. 5.8 in terms of the weight fraction

$${}^w\gamma_p^\infty = \frac{\rho_L RT}{K_{fg}p_1^0 M_1} \tag{5.14}$$

Clearly, the equation still maintains its form except that now the solute's molecular weight is used instead of the stationary phase. The weight fraction infinite-dilution activity coefficient, ${}^w\gamma_1^\infty$, can be expressed by rewriting eq. 5.10 as

$$\ln {}^w\gamma_1^\infty = \ln {}^w\gamma_p^\infty - \frac{p_1^0(B_{11} - \nu_1^0)}{RT} = \ln \frac{\rho_L RT}{K_{fg}p_1^0 M_1} - \frac{p_1^0(B_{11} - \nu_1^0)}{RT} \tag{5.15}$$

For the GC method, K_{fg} values can be obtained from the following equation

$$K_{fg} = (t_R - T_A)\left[F\left(\frac{T}{T_m}\right)\left(\frac{p_m - p_w}{p_m}\right)\right]\frac{3}{2}\left[\frac{(p_i/p_o)^2 - 1}{(p_i/p_o)^3 - 1}\right]\frac{1}{V_L} \tag{5.16}$$

where t_R and t_A are the retention time of the solute and a nonsorbed compound, respectively; F is the column flow measured by a soap-bubble flow meter; T and T_m are the temperatures of the column and the flow meter; p_m and p_w are the flow meter pressure and the saturated water vapor pressure; p_i and p_o are the inlet and outlet pressures of the column; V_f is the volume of the column stationary phase. Usually, p_m and p_o are equal to atmosphere pressure. By measuring the column flow rate F and retention times t_R and t_A, we can obtain K_{fg} values for solutes.

For the SPME method, the measurement of K_{fg} is straightforward. A gas sample is prepared by spiking n_0 mass of a solute into a capped vial with a capacity of V_g. Then, SPME sampling is performed until equilibrium is reached between the coating and the gas phase. If the volume of the vial is sufficiently large, the concentration change of the solute in the vial before and after SPME is negligible. The partition coefficient of the solute between the coating and the gas phase can be calculated from

$$K_{fg} = \frac{C_f}{C_g} = \frac{n_f}{V_f}\frac{V_g}{n_0} \tag{5.17}$$

where C_f^∞ and C_g^∞ are the solute concentrations in the coating and the gas phase at equilibrium, respectively; n_f and n_0 are the mass extracted by the fiber coating and the mass spiked into the vial; V_f and V_g are the volumes of the coating and the vial. To ensure that the concentration of solute in the vial does not change during measurement, the vial which contains the gas mixture should have a volume about 10^6 times larger than the volume of the fiber coating. The mass extracted by the fiber can be quantified through a GC/MS. During the SPME measurements, the concentration of the solute in the fiber coating must be very low to ensure the validity of the assumption of infinite dilution.

In the GC method, the gas phase is helium carrier gas, whereas in the SPME method, the gas phase is air and some methanol introduced during spiking. However, this difference in the gas composition, as expected, did not produce a noticeable effect. The poly(dimethylsiloxane) coating has poor affinity for methanol, so the effect of a small quantity of this solvent on the partition process is negligible. The use of methanol can be eliminated by using standard gas mixtures produces by device described at the beginning of this chapter.

As Table 5.18 shows, precision for the GC method gets worse as the column temperature rises. This is because the retention time of an analyte decreases as the column temperature rises in isothermal gas chromatography. Although the standard deviation (SD) of the retention time does not change much as

Table 5.18 K_{fg} Values for Benzene, Standard Deviations, and Relative Standard Deviations from Seven Replicates Using the SPME and the GC Methods

SPME/GC/MS				GC/FID			
Temp. (°C)	K_{fg}	SD	% RSD	Temp. (°C)	K_{fg}	S.D.	% RSD
25	421	10	2.4	25	423.4	0.95	0.22
50	164	3.7	2.2	50	163.3	0.77	0.47
80	66	3.4	5.2	80	67.3	0.96	1.4
100	44	2.3	5.2	100	42.7	0.97	2.3

the column temperature increases, the relative standard deviation (RSD) increases due to the smaller values of the retention time. Overall, the GC method has better precision than SPME. For the latter method, the GC/MS quantitation and solvent injection used for detector callibration, each have a few percent of random error, which is difficult to minimize. In addition the column temperature can be more precisely controlled than that of the oil bath used for SPME. For the measurement of partition coefficients at low temperature, the precision of the GC method is undoubtedly better. As temperature increases, the difference in precision between the two methods is reduced. The K_{fg} values are higher in Table 5.18 compared to Table 5.1 because lower cross-linked PDMS was used.

The distribution constant can then be used to calculate the activity coefficients for probe molecules in appropriate coatings.[4] Table 5.19 summarizes the infinite dilution weight fraction activity coefficients (ln ${}^{w}\gamma_{p}^{\infty}$) and their virial corrected counterparts (ln ${}^{w}\gamma_{1}^{\infty}$) of the five McReynolds solutes measured by SPME in two polymeric coating materials: one is 7 μm poly(dimethylsiloxane) bonded phase (equivalent to SPB-1), the other is 15 μm 50% phenyl substituted poly(dimethylsiloxane) (equivalent to SPB-50). The weight-fraction infinite dilution activity coefficients of the five probe compounds are indicators of how these compounds interact with the coating materials, thus they can be used to characterize the coating's properties. The ${}^{w}\gamma_{p}^{\infty}$ values of the five compounds in the 50% phenyl poly(dimethylsiloxane) are quite different from those in poly(dimethylsiloxane). The most obvious change is the substantial decrease of ln ${}^{w}\gamma_{p}^{\infty}$ values for benzene and pyridine. Compared to the poly(dimethylsiloxane) coating, the 50% phenyl coating (SPB-50) provides a physicochemical "environment" more similar to its own molecular structure because of the large number of phenyl groups. Since the structural differences between the solutes and the solvent become smaller, the ln ${}^{w}\gamma_{p}^{\infty}$ values also become smaller.

The initial results indicate that SPME is suitable for the characterization of coating materials. Some improvements need to be made to take full advantage of the method. One is the precision, which is relatively poor when compared with the GC technique. This can be improved by more careful and precise control of sampling temperature and by using an internal standard to

Table 5.19 Weight Fraction Infinite Dilution Activity Coefficients of Five Probe Compounds (${}^{w}\gamma_p^{\infty}$) and Their Virial Corrected Counterparts (${}^{w}\gamma_1^{\infty}$) in Two Coatings at Four Different Temperatures

		SPB-1		SPB-50	
	T(°C)	$\ln {}^{w}\gamma_p^{\infty}$	$\ln {}^{w}\gamma_1^{\infty}$	$\ln {}^{w}\gamma_p^{\infty}$	$\ln {}^{w}\gamma_1^{\infty}$
Benzene	25	1.71	1.72	0.56	0.57
	50	1.69	1.71	0.81	0.82
	80	1.57	1.6	0.46	0.5
	100	1.53	1.58	0.55	0.61
1-Butanol	25	3.66	3.66	3.07	3.07
	50	3.2	3.2	2.79	2.8
	80	2.59	2.6	1.93	1.94
	100	2.48	2.5	1.72	1.74
2-Pentanone	25	2.19	2.2	0.99	1
	50	2.03	2.05	1.04	1.05
	80	1.85	1.88	1.02	1.04
	100	1.74	1.78	1.03	1.07
Nitropropane	25	2.66	*	1.08	*
	50	2.55	*	1.22	*
	80	2	*	0.89	*
	100	1.96	*	0.86	*
Pyridine	25	1.29	1.29	0.34	0.34
	50	1.21	1.21	0.6	0.61
	80	1.91	1.92	0.32	0.33
	100	1.64	1.66	0.39	0.41

* Cannot be calculated because ν_1^0 and B_{11} values are unavailable.

minimize the concentration variation due to imprecise spiking. Another possible improvement is to narrow the injection band of probe molecules from the fiber desorption. This will allow a GC/MS detection of lower solutes quantities, thus solute concentrations in the coating may be further lowered to ensure the assumption of infinite dilution. The combination of better precision and improved detection limits will achieve accurate determination of infinite dilution activity coefficients and establish SPME as a viable alternative to the GC method for the characterization of stationary phases.

5.8.2 Partitioning in Multiphase Matrices

The investigation of chemical species distribution in natural systems is critical for many disciplines of science including environmental chemistry. The information gained during these studies is also very important to the analytical chemist, since it facilitate development of optimum sampling protocols and can lead to a more rational optimization of extraction parameters. One approach to the study of multiphase equilibria involves the separation of individual phases after equilibrium has been achieved, and then to analyse them individually to determine the concentration of target compounds. For example, SPME was used to study the concentration of organic substances from octanol-saturated water.[52] However, the separation of phases is not necessary, as illustrated below in an example describing a practical multiphase system involving humic material in aqueous matrix.

In a sample with several matrix components (true phases or pseudophases, e.g., colloidally dissolved organic matter—DOM) $i = 1, 2, 3 \ldots$ the nominal concentration of a target analyte m in a component i can be written

$$C_{i,m} = \frac{n_m}{V_i + \sum_{j \neq i} V_j K_{i,j,m}} \tag{5.18}$$

where n_m is total amount of analyte in the sample, $K_{ij,m}$ is distribution constant of a target analyte, m, between the components i and j ($K_{ij,,m} = C_{i,m}/C_{j,m}$ at equilibrium), and V_i is volume of component i. Volume can be replaced by mass or surface area if a sample component is better characterized in those terms, with $K_{ij,m}$ in appropriate units.

Suppose a sample consists of water and one other pseudophase, such as dissolved organic matter (DOM). With the fiber not present, the concentration of the free analyte m in water w at equilibrium would be:

$$C_{m,w} = \frac{n_m}{V_W + m_{\text{DOM}} K_{\text{DOMw},m}} \tag{5.19}$$

assuming that the analyte concentration in headspace is negligible, where V_W is the volume of the water phase, m_{DOM} the weight of DOM, n_m is in grams, and $K_{\text{DOMw},m}$ the partition coefficient, defined as $K_{\text{DOMw},m} = C_{\text{DOM},m}/C_{w,m}$ and expressed in mL/g units, when $C_{\text{DOM},m}$ is in g/g and $C_{w,m}$ is in g/mL. $C_{\text{DOM},m}$ and $C_{w,m}$ are the concentrations of analyte, m, in DOM and water respectively. With the fiber coating present in the system, the free analyte concentration at equilibrium would be:

$$C'_{w,m} = \frac{n_m}{V_w + m_{\text{DOM}} K_{\text{DOMw},m} + V_f K_{fwm}} \tag{5.20}$$

where V_f is the fiber coating volume and $K_{fw,m}$ the coating/water partition coefficient for component m. Thus the ratio of the concentration of the free target analyte, m, in water before SPME, $C_{m,w}$, to that after SPME, $C'_{w,m}$ is:

$$\frac{C_{w,m}}{C'_{w,m}} = 1 + \frac{V_f K_{fwm}}{V_w + m_{\text{DOM}} K_{\text{DOMw},m}} \tag{5.21}$$

For the concentration of the free target analyte in water to change by less than 10%, the criterion is

$$\frac{V_f K_{fwm}}{V_w + m_{\text{DOM}} K_{\text{DOMw},m}} < 0.111 \tag{5.22}$$

If the value of the sorption term $m_{DOM}K_{DOMw,m}$ is unknown, this criterion is safely stated as:

$$K_{fw,m} < \frac{V_w}{10V_f} \quad (5.23)$$

or $V_w(\mathrm{mL}) > (K_{fw,m}/164)$ for a fiber with a 100 μm thick coating, and $V_w(\mathrm{mL}) > K_{fw,m}/3887$ for a 7 μm thick coating. The value of $K_{fw,m}$ can be determined by performing SPME on samples of deionized pure water spiked with analyte standards. Assuming an SPME vial of 40 mL, the $K_{fw,m}$ of the analyte m for a 7 μm fiber must not be greater than 155,500 in order to avoid a disturbance of the partitioning equilibrium by the fiber uptake by more than 10%. This condition is met even for highly hydrophobic PAHs. Thus SPME can extract analytes without significantly changing their concentration in a sufficiently large volume of water. Hence, SPME can determine the concentration of free analytes dissolved in water in a DOM/water sample.

Equation 5.19 can be rearranged to eq. 5.24 such that the correlation between the concentrations of free dissolved analyte m, $C_{w,m}$, reversibly sorbed portion, $C_{DOM,m}$ and its total concentration, $C_{total,m}$ is explicitly expressed as:

$$C_{total,m} = C_{w,m} + C_{DOM,m} = C_{w,m}(1 + K_{DOMw,m}m_{DOM}/V_w) \quad (5.24)$$

where V_w is the matix volume, which corresponds closely to volume of the water for low DOM concentration. On the basis of eq. 5.25, the total concentration of an analyte $C_{total,m}$ in a sample, i.e., the sum of the freely dissolved and the reversibly bound portion, can be determined by SPME using isotopically labeled internal standards (st):

$$C_{total,m} = C_{total,st}\frac{GC_m}{GC_{st}} \quad (5.25)$$

where GC_m and GC_{st} are the GC responses after SPME of the nonlabeled analyte (native) and the labeled standard (spike), respectively. The total concentration of the deuterated standard is a known value. Nondeuterated analytes and deuterated surrogates have identical partition coefficients. Another prerequisite is that the partitioning equilibrium be established rapidly (in order to perform experiments in a reasonable time). Alternatively, total concentrations can be determined by other methods, such as standard addition or exhaustive liquid-liquid extraction (LLE).

In contrast to internal calibration by means of deuterated surrogates, conventional external calibration using spiked pure water samples gives the free dissolved concentration. Knowledge of the total concentrations (either from internal calibration or from LLE) and free dissolved fractions allow calculation of partition coefficients K_{DOM}. Table 5.20 summarizes results obtained with SPME versus LLE for total concentration determination and literature data.[53]

Table 5.20 Concentrations of Pollutants (phenols in ppm, PAHs in ppb) in a Wastewater: Sample Analyzed by LLE, SPME; Calculations According to eq. 5.24 and Literature Data

Analyte, m	LLE (Internal Standards)	(% RSD)	SPME	Calculated	SPME (External Calibration[a])	(% RSD)	log $K_{DOM,w}$[54]
Phenol	104.	(8.6)	94.	97.3	97.	(5.3)	0.88
o-Cresol	6.0	(7.4)	6.0	6.2	6.2	(3.7)	1.35
m/*p*-Cresol	34.5	(7.2)	[b]	35.9	35.6	(3.8)	1.35
Naphthalene	25.7	(3.8)	27.5	26.1	21.6	(5.9)	2.79
Fluorene	7.8	(4.0)	7.3	7.7	3.4	(6.6)	3.58
Phenanthrene	12.3	(6.1)	[b]	8.9	2.1		3.98
Anthracene	3.4	(6.3)	3.5	3.0	0.55		4.11
Fluoranthene	4.9	(12.4)	[b]	4.1	0.40		4.44
Pyrene	5.2	(10.3)	4.9	4.4	0.35		4.53
Chrysene	2.4	(16.1)	1.8	1.3	0.2		[b]

[a] Concentrations of freely dissolved pollutants, in contrast to columns 2 and 3, which give total concentrations in the matrix.
[b] Deuterated surrogate was not available.

SPME provides good agreement with standard techniques which indicates that it is well suited for such investigations.

The technique to measure analyte partition coefficients between two components of a sample can be extended to aqueous suspensions, such as sludge and many other systems, for example, air and air particulate, or even water mist (fog) in air. The method uses simple and commercially available apparatus, is cost and time efficient, and SPME has been shown to be sensitive to many analytes in air and water.

5.8.3 Kinetics of the Partitioning Process

Partitioning kinetics is another important issue for analytical chemists, since it determines extraction rates. Interaction of nonpolar organic analytes with dissolved polymers is said to be fast because physisorption does not require high activation energies. SPME can be used to investigate the kinetics of the equilibration process of organic pollutants between water and dissolved HOM. In such studies the sample containing native analytes and HOM is spiked with isotopically labeled analogues. Rapid extraction from a stirred HOM/water sample can monitor concentration changes of target analytes with respect to time. Figure 5.27 illustrates the changes in concentration of labeled compound in water versus native analyte as a function of time. The curves indicates the rapid equilibration of the species between water and HOM. The data presented in Figure 5.27 do not rule out, however that there may be additional sorption process with a much lower rate.

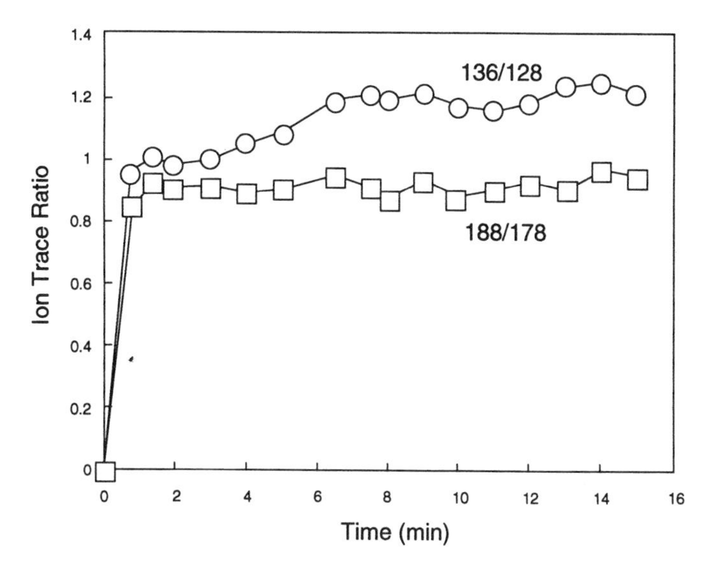

Figure 5.27 Kinetics of the sorption of naphthalene and anthracene onto DOM (100 ppm) from water measured by SPME; *y* axis represents ratio of GC signals for deuterated spike and nondeuterated compound. Matrix: diluted wastewater with native PAHs.

References

1. J. Adkins and N. Henry III *Anal. Chem.* ***65***, 133R (1993).
2. M. Chai and J. Pawliszyn *Environ. Sci. Technol.* ***29***, 693 (1995).
3. P. Martos and J. Pawliszyn *Anal. Chem.* ***69***, 206 (1997).
4. Z. Zhang and J. Pawliszyn *J. Phys. Chem.* ***100***, 17648 (1996).
5. The Sadtler Capillary GC Standard Retention Index Library and Data Base, Sadtler Research Laboratories, Philadelphia.
6. P. Martos, A. Saraullo and J. Pawliszyn *Anal. Chem.* ***69***, 402 (1997).
7. L. Pan, M. Adams and J. Pawliszyn *Anal. Chem.* ***67***, 4396 (1995).
8. A. Saraullo, P. Martos and J. Pawliszyn *Anal. Chem.*, in press.
9. C. Arthur, L. Killam, S. Motlagh, M. Lim, D. Potter and J. Pawliszyn *Environ. Sci. Technol.* ***26***, 979 (1992).
10. J. Dean, W. Tomllnson, V. Makovskaya, R. Cumming, M. Hetheridge and M. Comber *Anal. Chem.* ***68***, 130 (1996).
11. C. Arthur, L. Killam, K. Buchholz and J. Pawliszyn *Anal. Chem.* ***64***, 1960 (1992).
12. Z. Zhang and J. Pawliszyn *Anal. Chem.* ***67***, 34 (1995).
13. A. Boyd-Boland and J. Pawliszyn *J. Chromatogr.* ***704***, 163 (1995).
14. B. MacGillivray, *Analysis of Substituted Benzenes in Environmental Samples by Headspace Solid Phase Microextraction*, M.Sc. Thesis, (University of Waterloo, Waterloo, 1996).

15. A. Saraullo, *Determination of Petroleum Hydrocarbons in the Environment by Solid Phase Microextraction*, M.Sc. Thesis (University of Waterloo, 1996).

16. B. MacGillivray, P. Fowlie, C. Sagara and J. Pawliszyn *J. Chromatogr. Sci.* ***32***, 317 (1994).

17. T. Nilsson, F. Pelusio, L. Montanarella, B. Larsen, S. Facchetti and J. Madsen *J. High Resolut. Chromatogr.* ***18***, 617 (1995).

18. T. Gorecki, P. Martos and J. Pawliszyn *Anal. Chem.*, submitted.

19. D. Potter and J. Pawliszyn *Environ. Sci. Technol.* ***28***, 298 (1994).

20. J. Langenfeld, S. Hawthorne and D. Miller *Anal. Chem.* ***68***, 144 (1996).

21. K. Buchholz and J. Pawliszyn *Anal. Chem.* ***66***, 160 (1994)

22. L. Pan and J. Pawliszyn *Anal. Chem.* ***69***, 196 (1997).

23. L. Pan and J. Pawliszyn *J. Chromatogr.,* in press.

24. S. Johansen and J. Pawliszyn *J. High Resolut. Chromatogr.* ***19***, 627 (1996).

25. R. Eisert and K. Levsen *J. Am. Soc. Mass Spectrom.* ***6***, 1119 (1995).

26. I. Barnabas, J. Dean, I. Fowlis and S. Owen *J. Chromatogr.* ***705***, 305 (1995).

27. K. Graham, L. Sarna, G. Webster, J. Graynor and H. Ng *J. Chromatogr.* ***725***, 129 (1996).

28. A. Boyd-Boland, S. Magdic and J. Pawliszyn *Analyst* ***121***, 929 (1996).

29. P. Popp, K. Kalbitz, G. Oppermann *J. Chromatogr.* ***687***, 133 (1994).

30. Y. Cai and J. Bayona *J. Chromatogr.* **696**, 113 (1995).

31. T. Gorecki and J. Pawliszyn *Anal. Chem.*

32. T. Gorecki, H. Yuan and J. Pawliszyn, *in preparation.*

33. A. Boyd-Boland and J. Pawliszyn *Anal. Chem.* ***68***, 1521 (1996).

34. S. Scypinski, A-M Smith, L. Nelson and S. Shaw, private communication.

35. D. Page and G. Lacroix *J. Chromatogr.* ***648***, 199 (1993).

36. X. Yang and T. Peppard *J. Agric. Food Chem.* ***42***, 1925 (1994).

37. F. Pelusio, T. Nilsson, L. Montanarella, R. Tilio, B. Larsen, S. Facchetti and J. Madsen *J. Agric. Food Chem.* ***43***, 2138 (1995).

38. A. Steffen and J. Pawliszyn *J. Agric. Food Chem.* ***68***, 3008 (1996).

39. V. Mani and C. Woolley *LC-GC.* ***13***, 734 (1995).

40. S. Hawthorne, D. Miller, C. Arthur and J. Pawliszyn *J. Chromatogr.* ***603***, 185 (1992).

41. C. Grote and J. Pawliszyn *Anal. Chem.* ***69***, 587 (1997).

42. N. Nagasawa, M. Yashiki, Y. Iwasaki, K. Hara and T. Kojima *Forensic Sci. Int.* ***78***, 185 (1996).

43. L. Wang *Determination of Amphetamine and Methamphetamine in Urine and Blood Plasma by Headspace Solid Phase Microextraction*, M.Sc. Thesis, (University of Waterloo, Waterloo, Canada, 1996).

44. M. Krogh, K. Johansen, F. Tonnesen and K. Rasmussen *J. Chromatogr.* ***673***, 299 (1995).

45. T. Kumazawa, X-P Lee, M-C Tsai, H. Seno, A. Ishii and K. Sato *Jpn. J. Forensic Toxucol.* ***13***, 25 (1996).

46. K.G Furton and J. Bruna *J. High Resolut. Chromatogr.* ***18***, 625 (1995).

47. A. Steffen and J. Pawliszyn *Anal. Com.* ***33***, 129 (1996).

48. E. Otu and J. Pawliszyn *Microchim. Acta* ***112***, 41 (1993).

49. J. Chongrong and J. Pawliszyn, *J. Microcolumn. Sep.* submitted.

50. J-L Liao, C-M Zeng, S. Hjerten and J. Pawliszyn *J. Microcolumn. Sep.* ***8***, 1 (1996).

51. R. Laud and R. Pecsok *Physicochemical Applications of Gas Chromatography* (John Wiley and Sons, New York, 1978).

52. P. Popp, A. Paschke, U. Schroter, G. Oppermann *Chemia Analityczna.* ***40***, 897 (1995).

53. J. Poerschmann, Z. Zhang, F. Kopinke and J. Pawliszyn *Anal. Chem.* ***69***, 597 (1997).

54. J. Poerschmann, F. Kopinke and U. Stottmeister *Environ. Sci. Technol.* ***29***, 941 (1995).

CHAPTER

6

Experiments

The main objective of this chapter is to make available to the reader laboratory experiments that demonstrate the basic principles of the SPME technique. These exercises provide an opportunity for scientists who have not used this new technology to become familiar with it.

6.1 Determination of BTEX in Water

Goals of the Experiment

This set of experiments illustrates the stages of method development for new SPME applications. It also demonstrates some of the applicable quantitation techniques.

Materials and Equipment

- 100 ppm solution of BTEX (benzene, toluene, ethylbenzene, *o*-xylene) in methanol
- Aqueous solution of BTEX of unknown concentration
- SPME holder with 100 μm PDMS fiber
- 40 mL amber glass vials with 1 in. Teflon-coated stir bars (11), containing 25 mL of NANOpure (or equivalent) water each
- Magnetic stirrer
- 100 μL syringe

- Gas chromatograph equipped with a 30 m × 0.25 mm × 1 μm SPB-5 column and an FID
- Gas chromatographic conditions: injector temperature, 250°C; desorption time, 1 minute; oven temperature program: 70°C for 1 minute followed by 30°/min ramp to 180°C

Determination of the Desorption Time

Desorption time depends on several parameters, the most important of which are analyte volatility, injector design, and temperature. The effect of volatility is straightforward: the more volatile the compound, the faster the desorption.

Injector design can affect the desorption process in several ways. For example, temperature distribution in the injector may be nonuniform. It is important to position the fiber in the injector in a way that subjects it to nominal injector temperature. A lower temperature may cause unnecessarily slow desorption, while an excessively high temperature may destroy the coating. It is therefore advantageous to know the actual temperature profile of the injector. Such information usually can be found in the injector documentation. If not, the profile can be determined by using a small diameter thermocouple; alternatively, several desorptions can be performed at different depths to find optimal positioning of the fiber.

Of even greater importance is the internal volume of the injector (or, more precisely, its insert), which determines the linear flow rate of the carrier gas along the fiber. This flow rate should be as high as possible for efficient desorption; therefore the internal volume of the insert should be as small as possible. The septum programmable injector (SPI) offered by Varian is equipped with inserts having internal volume of a few microliters, which is ideal for SPME. For split/splitless injectors, Supelco offers special SPME inserts of 0.75 mm i.d. which are strongly recommended. If they are not available, splitless inserts should be used.

Increased temperature shifts the gas/coating partition coefficient of the analyte in favor of the gaseous phase. Consequently, the higher the temperature, the faster the desorption. However, high temperatures cause increased fiber background levels and may lead to accelerated coating deterioration.

The following experiment compares two desorption times. In real method development, it is customary to examine three or four different desorption times.

Procedure

1. Prepare a 100 ppb solution of BTEX in water by adding 25 μL of the standard methanolic solution to a vial equipped with a stir bar and prefilled with 25 mL of water.
2. Place the vial on a magnetic stirrer and mount it in a clamp.
3. Set the revolutions to 1200 rpm.

4. Pierce the septum of the vial with the needle of the SPME device, and fasten the device with a clamp. The needle tip should be positioned about 1 cm below the septum.
5. Expose the fiber by depressing the plunger, and lock it in the bottom position by turning it clockwise. The fiber should be located in the headspace above the sample.
6. Start the timer. The extraction should last 2 minutes.
7. Withdraw the fiber back into the needle and pull the device out of the vial.
8. Introduce the needle of the SPME device to the injector of the GC. Start the analysis by depressing the plunger and locking it in the "lowered" position.
9. After exactly 30 seconds withdraw the fiber back into the needle, and pull the needle out of the injector.
10. When the separation is completed, repeat the analysis to determine fiber carryover.
11. Repeat steps 4–10 (using the same standard solution) using exactly 1 minute desorption time.
12. Based on the carryover from the two analyses, chose the desorption time to be used throughout the rest of the experiment.

Determination of Extraction Time Profile

The amount of time necessary for the extraction process to reach equilibrium depends on numerous factors, the most important of which are mass transport conditions, nature of the analyte, and temperature. In general, extraction directly from the sample takes the longest when the sample is unstirred, thus subject only to natural convection. In this case, the analyte molecules can reach the fiber coating only via diffusion, which in liquids is a relatively slow process. Perfect agitation conditions are unobtainable by means of common methods, but any kind of stirring will usually shorten the equilibration time. Headspace sampling is advantageous for volatile analytes, because equilibrium can be reached much faster, provided the stirring is as vigorous as possible to facilitate analyte transfer from the liquid to the gaseous phase, or that the analytes are already in the gas phase at the beginning of the experiment.

The nature of the analyte can affect the equilibration time in many ways. Volatile analytes usually are extracted faster than semi- and nonvolatile ones, owing to higher diffusion coefficients. The equilibration time is usually shorter for analytes of lower sample/coating partition coefficients, since less analyte has to get to the coating through the stationary aqueous layer to reach equilibrium (e.g. benzene vs. xylene). When sampling from headspace, analytes that have small Henry's constants (are well soluble in water) tend to equilibrate more slowly, since at any given moment only a limited number of analyte molecules can be transported through the headspace which acts as a bottleneck (e.g. benzene vs. ethanol).

Temperature increases the diffusion coefficients of the analyte molecules and decreases the analyte/coating partition coefficients; hence equilibration times are shorter at elevated temperatures. On the other hand, increasing the temperature adversely affects sensitivity and complicates the experimental setup.

In the following experiment, the extraction time profile is determined at room temperature and at optimal stirring conditions only.

Procedure

1. Prepare a 100 ppb solution of BTEX in water by adding 25 μL of standard methanolic solution to a vial equipped with a stir bar and prefilled with 25 mL of water.
2. Place the vial on a magnetic stirrer and mount it in a clamp.
3. Set the revolutions to 1200 rpm.
4. Pierce the septum with the needle, mount the device and start the extraction as in the preceding experimental procedure.
5. Start the timer. The extraction should last exactly 1 minute.
6. Withdraw the fiber back into the needle and pull the device out of the vial.
7. Introduce the needle of the SPME device to the injector of the GC. Start the analysis by depressing the plunger and locking it in the "lowered" position.
8. After the desorption time determined in the preceding experiment, withdraw the fiber back into the needle, and pull the needle out of the injector.
9. Prepare a fresh 100 ppb solution of BTEX in water by adding 25 μL of standard methanolic solution to a new vial equipped with a stir bar and prefilled with 25 mL of water.
10. When the separation is completed, repeat steps 2–9 for extraction times of 2, 4, and 8 minutes.
11. Plot the results (peak area vs. extraction time) and determine the equilibration time to be used in all other experiments.

Determination of Method Precision

SPME is an equilibrium method, therefore precision of the results can affect overall method accuracy. It is very important that all possible sources of poor precision be identified and eliminated during method development. Some of the more frequent sources of poor precision are as follows:

- **Inconsistent stirring.** Digitally controlled magnetic stirrers are recommended. If they are not available, all experiments should be performed at the same stirrer setting. It should be kept in mind, however, that the stirring rate of stirrers without digital control can be affected by many factors, including line voltage and temperature.

- **Variable temperature.** In the most unfavorable case, the laboratory air temperature will vary significantly during the day. The use of a thermostated water bath is recommended in such cases. Also, some stirrers tend to heat up, especially at high revolutions. A warm stir plate can change the temperature of the sample during extraction, thus increasing the spread of the results. The use of a higher quality stirrer is recommended in such cases. A quick, but not ultimate solution is to place the vial about 0.5 cm above the stir plate, to prevent direct heating.
- **Inconsistent extraction time.** Care should be taken to extract both standards and unknown samples for exactly the same time. If equilibrium is normally reached, small variations of the extraction time will not significantly affect the precision. If, however, the extraction time selected is too short for the analytes to reach equilibrium, small variations in the extraction time can cause significant deterioration in precision.
- **Inconsistent desorption time.** Overly long desorption times usually do not affect precision, but desorption times that are too short result in incomplete desorption of the analyte.
- **Too small sample volume.** In the case of analytes characterized by large partition coefficients, a large percentage of the analyte is extracted by the fiber, especially from small sample volumes. Small variations in sample volume or analyte concentration may lead to significant differences in the amount extracted.
- **Inaccurate preparation of standards or spikes.**
- **Analyte losses due to poor selection of materials contacting the sample** (details explained in the following section).

The precision of a method should be estimated on the basis of a series of seven replicate sample analyses.

Determination of the Calibration Curve (Linear Range of the Response)

It follows from the theory of SPME (see Chapter 3) that the amount of analyte extracted from the sample under set conditions should always be directly proportional to the concentration of the analyte in the sample. It is obvious, therefore, that a linear calibration curve is normally expected. Deviations from linearity that are occasionally observed are in most cases not related to the extraction step itself.

At very low concentrations, deviations from linearity are usually due to analyte losses before and during the extraction process. Adsorption, absorption, and permeation are the main phenomena responsible for the losses. Proper choice of the sample storage container is of utmost importance. Glass vials are the containers of choice because of their low price and high inertness.

When polar compounds are analyzed, however, a significant portion of the analyte may be lost owing to adsorption on glass walls. In such cases, silanization of the vials can greatly improve the results. Transparent Teflon vials can be used as well, but they are very expensive. Also, losses of volatile analytes can occur in Teflon vials, especially after prolonged storage, as a result of permeation.

Extreme care should also be taken when choosing the materials that will contact the sample. Only Teflon- or glass-coated stir bars should be used. Vial septa must be Teflon coated. Any exposed area of silicone rubber absorbs analytes in exactly the same way as the fiber coating!

At high concentrations, deviations from linearity have different sources. Very often the amount of analyte extracted by the fiber will exceed the linear range of the detector. Only a flame ionization detector (FID) has a linear range spanning over seven orders of magnitude. The linear range of all other common GC detectors is much narrower, or even nonexistent (as in flame photometric detection).

Another possible reason for deviations from linearity at high concentrations is limited analyte solubility. This should be checked before the high concentration standard is prepared. Even if theoretically the solubility of the analyte in water would not be exceeded under the conditions contemplated, it should be kept in mind that the kinetics of the dissolution process can be extremely slow (especially if the aqueous solubility of the analyte is low).

In the following experiment the calibration curve will be determined for four concentrations: 50, 100, 200, and 400 ppb.

Procedure

1. Prepare a 50 ppb solution of BTEX in water by adding 12.5 μL of standard methanolic solution to a vial equipped with a stir bar and prefilled with 25 mL of water.
2. Place the vial on a magnetic stirrer and mount it in a clamp.
3. Set the revolutions to 1200 rpm.
4. Pierce the septum with the needle, mount the device and start the extraction. Extraction time determined in the earlier experiment should be used.
5. When the extraction is finished, perform the GC analysis as in the preceding experiments.
6. Prepare a 100 ppb solution of BTEX in water by adding 25 μL of standard methanolic solution to a new vial equipped with a stir bar and prefilled with 25 mL of water.
7. Repeat steps 2–6 for 200 and 400 ppb solutions (50 and 100 μL of the methanolic standard should be added to water, respectively).
8. Plot the results (peak area vs. concentration) and use linear regression to determine the equation of the line and the linear correlation coefficient.

Quantitative Analysis

The choice of the quantitation method in SPME depends primarily on the nature of the sample matrix. Simple matrices (e.g., drinking water) do not interfere with the extraction process; hence a basic form of calibration curve quantitation can be used for both direct and headspace sampling. In more complex matrices, when some of matrix components (e.g., suspended matter) are liable to compete with the fiber in the extraction process or to modify the properties of the coating (e.g., surfactants), quantitation methods that take the matrix into account should be used. The simplest method of this kind is standard addition. Multipoint standard addition can yield more reliable results than single-point standard addition, yet it requires more determinations per sample. In any case, unless only a few determinations are performed, more analyses must be performed per sample than are needed in the calibration curve method. Standard addition can be used with any detector.

When a mass spectrometer is used for quantitation, very good results can be obtained by isotopic dilution. In principle, this is a combination of the internal standard and standard addition methods, in which the standard added is isotopically labeled. It is assumed that isotopic substitution does not significantly affect the properties of the compound, which is extracted in exactly the same way as the native analyte. The two can be differentiated during quantitation by the characteristic masses of the fragments formed. Only one analysis is required per sample to obtain quantitative results. Many possible sources of error (e.g., variable temperature, inconsistent stirring rate, inconsistent desorption time, etc.) are naturally accounted for, as both the analyte and the isotopically labeled analogue are extracted and analyzed during the same run, hence under the same conditions. The only drawbacks of this method are the high price and limited availability of the isotopically labeled compounds.

This experiment uses and compares two quantitation methods: external calibration curve and standard addition.

Procedure

1. Prepare an "unknown" solution of BTEX in water by adding 10–25 μL of the methanolic solution to a vial equipped with a stir bar and prefilled with 25 mL of water.
2. Place the vial on a magnetic stirrer and mount it in a clamp.
3. Set the revolutions to 1200 rpm.
4. Pierce the septum with the needle, mount the device, and start the extraction. Extraction time determined earlier should be used.
5. When the extraction is finished, perform the GC analysis.
6. Calculate the concentrations of the individual BTEX components in the

aqueous sample from the calibration curve determined in the preceding experiment.

7. Prepare another solution of BTEX in water by adding 25 μL of the methanolic solution of unknown concentration to a new vial equipped with a stir bar and prefilled with 25 mL of water, then add 25 μL of standard methanolic BTEX solution.
8. Repeat steps 2–5.
9. Calculate the concentrations of the individual BTEX components from standard addition data (peak areas of the components for the unknown sample and peak areas for the unknown sample with standard added, step 7).
10. Compare the results obtained by the two methods and establish the possible reasons for any discrepancies.

6.2 Determination of Pesticides in Water

Goals of the Experiment

The set of experiments illustrates various approaches to the analysis of samples containing trace levels of semivolatile analytes. Coating selection and matrix modification are illustrated. Quantitation is performed.

Materials and Equipment

- 10 ppm methanolic solution of pesticides (dichlorvos, EPTC, ethoprofos, trifluralin, simazine, propazine, diazinon, methyl chlorpyriphos, heptachlor, aldrin, metolachlor, endrin)
- Solution of pesticides of unknown concentration
- Sodium chloride
- SPME holder with PDMS-coated fiber (100 μm)
- SPME holder with poly(acrylate)-coated fiber
- 40 mL amber glass vials with 1 in. Teflon-coated stir bars (4), containing 25 mL of NANOpure (or equivalent) water each
- Magnetic stirrer
- 100 μL syringe
- Analytical balance
- Squeeze bottle with NANOpure (or equivalent) water
- Gas chromatograph equipped with a 30 m × 0.25 mm × 0.25 μm SPB-5 column, coupled to a mass spectrometer
- Set of library spectra of all the analytes
- Chromatographic conditions: injector temperature, 250°C; desorption time, 5 minutes; oven temperature program, 5 minutes at 40°C followed by ramps of 30°/min to 100°C, 5 deg/min to 250°C, 50°/min to 300°C, and then isothermal for 1 minute

Coating Selection

The pesticides contained in the standard solution represent all major groups: organochlorine, organonitrogen, and organophosphorus. Their physicochemical properties differ widely. Some of them are hydrophobic (e.g. organochlorine pesticides), while others are hydrophilic (e.g., organophosphates). It is therefore difficult to select a coating material that would be optimal for them all. In such cases, it is necessary to check the performance of different coatings and select the optimal one according to the initial goals set. The following criteria should be taken into account:

- **Amount extracted.** This will depend on the nature of the analyte and the coating. Nonpolar or slightly polar analytes are well extracted by PDMS coatings, while polar analytes are better extracted by poly(acrylate) coatings or Carbowax/DVB coatings. Usually the coating that produces the most uniform response for all the analytes should be selected.
- **Equilibration time.** This depends on a number of factors, including stirring conditions. If an analyte has a very large partition coefficient for a given coating, the number of analyte molecules that have to reach the coating is also large, and equilibration usually lasts longer.
- **Required precision.** This is related to the first two items. In the case of analytes characterized by large partition coefficients, a large percentage of the analyte is extracted by the fiber, especially from small sample volumes. Small variations in sample volume or analyte concentration may lead to significant differences in the amount extracted. Use of larger sample volumes can alleviate this problem. Also, if all analytes have reached equilibrium by the end of the extraction step, small variations of the extraction time will not affect the amount extracted. On the other hand, for increased sample throughput, it is often advantageous to set a shorter extraction time. In such cases, small variations in extraction time may translate into a significant spread of results.

Procedure

1. Prepare a 30 ppb solution of pesticides in water by adding 75 μL of the standard methanolic solution to a vial equipped with a stir bar and prefilled with 25 mL of water.
2. Place the vial on a magnetic stirrer and mount it in a clamp.
3. Set the revolutions to 800 rpm.
4. Pierce the septum of the vial off-center with the needle of the SPME device (PDMS fiber), and fasten the device with a clamp. The needle tip should be positioned directly above the sample, but to avoid wicking of the sample by capillary forces, it should not be immersed in it.
5. Expose the fiber by depressing the plunger, and turn the plunger clockwise to lock it in the bottom position. The fiber should be fully immersed in the sample.

6. Start the timer. The extraction should last 45 minutes.
7. Prepare a new 30 ppb pesticide solution as described in step 1.
8. Withdraw the fiber back into the needle and pull the device out of the vial.
9. Introduce the needle of the SPME device to the injector of the GC/MS system. Start the analysis by depressing the plunger and locking it in the "lowered" position.
10. Withdraw the fiber back into the needle after 3 minutes, and pull the needle out of the injector.
11. Repeat steps 2–6 and 8–10 for the poly(acrylate) fiber.
12. When the first separation is finished, identify all the analyte peaks by comparing their spectra with library spectra.
13. Prepare a quantitation file and integrate the peak areas using specific masses. Selection of the quantitation masses should be based on the analyte spectra.
14. When the second separation is finished, repeat steps 12 and 13.

Matrix Modification

There are several ways to enhance the extraction process in SPME. The amount extracted at equilibrium can be increased by reducing aqueous solubility of the analyte, increasing the concentration of the extractable form, or converting the analyte to a form that can be extracted more readily. The first two possibilities are usually based on matrix modification, while the third one requires derivatization of the analyte, which is outside the scope of this experiment. Partition coefficients are, in general, related to aqueous solubility of an analyte. The higher the solubility, the more the analyte tends to remain in the aqueous phase, and consequently the partition coefficient is lower. A simple way to change the solubility of an analyte through matrix modification is to add a large amount of a salt (usually NaCl). Aqueous solubilities of many organic compounds decrease in the presence of excess salt. On the other hand, for those for which the solubility does not change, the addition of salt may lead to a decrease in the amount extracted. This is related to the dependence of partition coefficients on activities rather than concentrations of the analytes in a solution. High salt content results in high ionic strength of the solution, which, in turn, may cause a significant decrease of the activity coefficients of some analytes.

Some analytes can be present in aqueous solutions in several forms (e.g., dissociated and nondissociated). Weak acids or bases constitute a good example. Normally only the nondissociated form can be extracted. The amount extracted at equilibrium therefore is liable to be significantly increased by a shift in the dissociation equilibrium toward the nondissociated form. Such a shift can be achieved by adding excess common ions. For acids or bases, this is done by pH modification. In general, weak acids will be much better extracted at low pH, while for weak bases a high pH is favored. It should be kept in mind, however, that addition of acid or base to the sample affects also the ionic strength, and thus the activity coefficients.

Only the effect of adding salt is studied in this experiment.

Procedure

1. Prepare a 30 ppb solution of pesticides in water by adding 75 μL of the standard methanolic solution to a vial equipped with a stir bar and prefilled with 25 mL of water.
2. Weigh 5 g of NaCl on the analytical balance and add it to the vial.
3. Place the vial on a magnetic stirrer and mount it in a clamp.
4. Set the revolutions to 800 rpm.
5. Pierce the septum of the vial off-centre with the needle of the SPME device (using the fiber selected in the preceding experiment), and fasten the device with a clamp. The needle tip should be positioned directly above the sample, but to avoid wicking of the sample by capillary forces, it should not be immersed in water.
6. Expose the fiber by depressing the plunger, and turn the plunger clockwise to lock it in the bottom position. The fiber should be fully immersed in the sample.
7. Start the timer. The extraction should last 45 minutes.
8. Withdraw the fiber back into the needle and pull the device out of the vial.
9. Expose the fiber by depressing the plunger and rinse it with a few milliliters of water from the squeeze bottle. This step aims at removing traces of NaCl from the coating surface. If the salt is not removed, it crystallizes on the fiber surface when it is exposed to the high temperature of the injector, which in turn adversely affects the precision of the results.
10. Introduce the needle of the SPME device to the injector of the GC/MS system. Start the analysis by depressing the plunger and locking it in the "lowered" position.
11. Withdraw the fiber back into the needle after 3 minutes, and pull the needle out of the injector.
12. Upon completion of the separation, integrate the peak areas using the quantitation file prepared previously.

Quantitative Analysis

The choice of a proper quantitation method for a given SPME analysis was discussed in detail in the previous experiment. Basic calibration curve quantitation can be used, since pesticide solution in pure water is analyzed in the current experiment. To save time, the procedure described below performs quantitative analysis based on single point calibration only. If time is not of concern, the full calibration curve can be developed. Selection of the fiber and the use of matrix modification should be decided based on the results of the preceding investigation.

Procedure

1. Prepare an "unknown" solution of pesticides in water by adding between 10 and 25 μL of standard methanolic solution to a vial equipped with a stir bar and prefilled with 25 mL of water.
2. Optional: If you choose to use matrix modification, weigh 5 g of NaCl on the analytical balance and add it to the vial.
3. Place the vial on a magnetic stirrer and mount it in a clamp.
4. Set the revolutions to 800 rpm.
5. Pierce the septum of the vial off center with the needle of the SPME device (using the fiber selected in the preceding experiments), and fasten the device with a clamp. The needle tip should be positioned directly above the sample, but to avoid wicking of the sample by capillary forces, it should not be immersed in water.
6. Expose the fiber by depressing the plunger and turn the plunger clockwise to lock it in the bottom position. The fiber should be fully immersed in the sample.
7. Start the timer. The extraction should last 45 minutes.
8. Withdraw the fiber back into the needle and pull the device out of the vial.
9. If salt was added, expose the fiber by depressing the plunger and rinse it with a few milliliters of water from the squeeze bottle.
10. Introduce the needle of the SPME device to the injector of the GC/MS system. Start the analysis by depressing the plunger and locking it in the "lowered" position.
11. Withdraw the fiber back into the needle after 3 minutes, and pull the needle out of the injector.
12. When the separation is completed, integrate the peak areas using the quantitation file prepared previously.
13. Calculate the concentrations of the analytes in the aqueous sample based on the results of one of the earlier experiments (depending on the conditions chosen).
14. Compare the results with the true values and discuss the possible reasons for any discrepancies.

6.3 Determination of Caffeine in Beverages

Goals of the Experiment

This experiment illustrates the practical aspects of SPME analysis of samples with complex matrices. Quantitation by isotopic dilution is also illustrated.

Materials and equipment

- 5 mg/mL solution of ^{12}C caffeine in methanol
- 5 mg/mL solution of $^{13}C_3$ (trimethyl) caffeine in methanol

- Freshly brewed tea, coffee, and decaffeinated coffee
- Canned cola-type soft drink
- SPME holder with uncoated fused silica fiber (uncoated fiber can be obtained by dissolving the PDMS coating in hot sulfuric acid)
- 15 mL glass vials with Teflon-coated stir bars (21)
- Magnetic stirrer
- 500 μL syringe
- Gas chromatograph equipped with a 30 m × 0.25 mm × 0.25 μm SPB-5 column and a mass selective detector
- Gas chromatographic conditions: injector temperature, 250°C; desorption time, 5 minutes; oven temperature program, 70°C for 1 minute followed by ramp at 30°/min to 220°C and then at 10°/min to 280°C.

Introduction

Caffeine naturally occurs in tea, coffee, and cola nuts. Caffeine analysis is performed for quality control purposes—for example, to verify that decaffeinated coffee contains no more caffeine than is permitted by regulatory standards. Current methods for the determination of caffeine in beverages require pH adjustments and often involve the use of toxic organic solvents. Alternatively, SPME can be used for this purpose. For extracting the caffeine from beverages, an uncoated fiber can be used, since caffeine contents in the beverages of interest are in the ppm range. Also, an uncoated fiber minimizes potential carryover problems and produces sharp injection bands owing to rapid desorption. The bare fiber can be prepared by placing a damaged coated fiber in nitric acid for a short time. Caution must be taken not to immerse the metal tubing holding the fiber. Alternatively, a 7 μm PDMS fiber can be used.

Beverage matrices can be very complex, and they vary from sample to sample. For samples of this nature, quantitation with external standards usually does not produce good results. The best results can be obtained by using the standard addition method. Application of GC/MS for the final determination makes it possible to use isotopically labeled analytes as internal standards. In such a case, the internal standard and the native analyte have the same chemical and physical properties, hence their behaviors during extraction (equilibration time, matrix effects) are identical. As a result, the use of an isotopically labeled internal standard essentially eliminates the need to reach equilibrium of analyte partitioning between the fiber and the sample. Also, any change in extraction conditions, including change of fiber properties due to irreversible adsorption of some of the matrix components, is compensated for. Quantitative information can be obtained by generating extracted mass chromatograms for specific ions of the native analyte and the isotopically labeled internal standard, and comparing the two peak areas obtained.

Determination of the Calibration Curve

1. Prepare two vials of 50 μg/mL caffeine solution in water by adding to each one 120 μL of the standard methanolic ^{12}C caffeine solution and 240 μL of $^{13}C_3$ caffeine solution (internal standard). Each vial should be prefilled with 12 mL of water and equipped with a stir bar.
2. Place the first vial on the magnetic stirrer and mount it in a clamp.
3. Set the revolutions to 800 rpm.
4. Pierce the septum with the needle, mount the SPME device in a clamp and start the extraction. Extraction time should be equal to exactly 5 minutes.
5. Withdraw the fiber back into the needle and pull the device out of the vial.
6. Introduce the needle of the SPME device to the injector of the GC held at 250°C. Start the analysis by depressing the plunger and locking it in the "lowered" position. Oven temperature program should be as follows: 1.5 minutes at 200°C, followed by a ramp of 30°C/min to 275°C.
7. After 1 minute desorption time withdraw the fiber back into the needle, and pull the needle out of the injector.
8. Repeat the analysis for the second vial.
9. Prepare two vials of 100 μg/mL caffeine solution in water by adding to each one 240 μL of the standard methanolic ^{12}C caffeine solution and 240 μL of $^{13}C_3$ caffeine solution. Each vial should be prefilled with 12 mL of water and equipped with a stir bar.
10. Repeat steps 2–8 for 200, 300, and 500 μg/mL solutions (480, 720, and 1200 μL of the methanolic standard, respectively, should be added to water. The volume of labeled caffeine solution added to each vial should be kept constant at 240 μL).
11. Determine the peak areas for the molecular ions of ^{12}C caffeine (m/z = 194) and $^{13}C_3$ caffeine (m/z = 197)
12. Plot the results (the ratios of peak areas m/z 194/197 vs. concentration) and use linear regression to determine the equation of the line and the linear correlation coefficient.

Analysis of Caffeine Content in Selected Beverages

1. Add 12 mL of one of the beverages and 240 μL of $^{13}C_3$ caffeine solution to a 15 mL vial equipped with a stir bar. Before proceeding, make sure that the sample is at room temperature.
2. Place the vial on the magnetic stirrer and mount it in a clamp.
3. Set the revolutions to 800 rpm.
4. Pierce the septum with the needle, mount the device, and start the extraction. Extraction time should be 5 minutes.
5. Upon completion of the extraction, perform the GC/MS analysis.
6. Repeat steps 1–5 to obtain duplicate results.
7. Repeat steps 1–6 for the remaining beverages.
8. Calculate the concentration of caffeine in each of the beverages from the

ratio of m/z 194/197 peak areas using the calibration curve determined in the preceding experiment.

Determination of Method Precision

1. Add 12 mL of the beverage analyzed last in the preceding experiment and 240 μL of $^{13}C_3$ caffeine solution to a 15 mL vial equipped with a stir bar. Before proceeding, make sure that the sample is at room temperature.
2. Place the vial on the magnetic stirrer and mount it in a clamp.
3. Set the revolutions to 800 rpm.
4. Pierce the septum with the needle, mount the device, and start the extraction. Extraction time should be 5 minutes.
5. Upon completion of the extraction, perform the GC/MS analysis.
6. Repeat steps 1–5 two more times for the same beverage.
7. Calculate the concentration of caffeine for each run using the same method as in the preceding experiment.
8. Estimate the precision of the method by calculating the relative standard deviation (% RSD) for the five results obtained for the same beverage (two from the caffeine content experiment and three from the present experiment).

APPENDIX

Theoretical Analyses

This appendix presents the complete mathematical derivations of theoretical results presented in the main text. This appendix is intended as a reference section and can be used when new applications of SPME require modifications to the theory. The simple geometry and processes of SPME permit refined theoretical analysis, but when formulae are too complicated to be practical approximations are derived based on the exact analytical solutions.

A.1 Estimation of the Distribution Constant for an Internally Cooled Fibre

The change of entropy (ΔS) in a system is determined by the initial and final state of the system and is independent of how the system changes from its initial to final state. As a result, we can calculate ΔS of a system by designing an artificial route, which has the same initial and final state, but makes calculation of ΔS easier. Figure A1.1 shows the initial and final state of the internally cooled SPME sampling and an alternative route indicated by (1), (2), and (3). ΔS of the cooled SPME sampling equals the sum of the entropy changes of all three steps in alternative route which can be expressed as:

$$\Delta S = \Delta S_1 + \Delta S_2 + \Delta S_3$$

In the following text, we calculate the entropy change in each step.

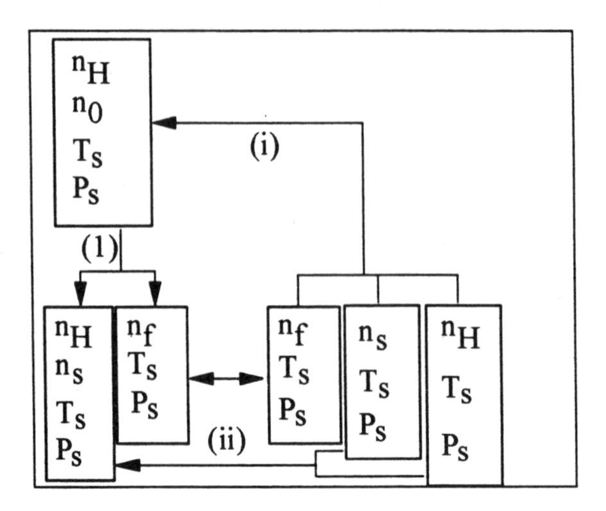

Figure A1.1 Alternative route for calculation of the Gibbs free energy change (DG_1) in Step 1; (i) mixing air (n_H) and all of the analyte vapor ($n_0 = n_s + n_f$), and (ii) mixing air (n_H) and part of the analyte vapor (n_s); two-headed arrow indicates no changes occur for that phase.

ΔS₁, Entropy Change in Step (1). Step (1) is a process under constant pressure and temperature. It is very simple to calculate the Gibbs free energy change (ΔG_1) in this process. ΔS_1 can be calculated from ΔG_1. We designed step (1) in which ΔG_1 can be calculated through $\Delta G_1 = \Delta G_{ii} - \Delta G_i$, where ΔG_i is due to mixing air (n_H) and all of the analyte vapor (n_0), and ΔG_{ii} is due to mixing air (n_H) and part of the analyte vapor (n_s). Keep in mind that $n_0 = n_f + n_s$, where n_0 is the total amount of the analyte, n_s is the amount of the analyte that is still mixed with air, and n_f is the amount of the analyte that is separated from the mixture. We should also point out that mixing n_s and n_f under constant temperature and pressure will not result in any change of the Gibbs free energy nor its entropy since the analyte molecules themselves are indistinguishable.

The ΔG for mixing two bases (n_1 and n_2) under constant temperature and pressure can be expressed as:

$$\Delta G = n_1 RT \ln \frac{n_1}{n_1 + n_2} + n_2 RT \ln \frac{n_2}{n_1 + n_2} \tag{A1.1}$$

From eq A1.1, we have ΔG_i:

$$\Delta G_i = N_H RT_s \ln \frac{n_H}{n_H + n_0} + n_0 RT_s \ln \frac{n_0}{n_H + n_0} \tag{A1.2}$$

and ΔG_{ii}:

$$\Delta G_{ii} = n_H RT_s \ln \frac{n_H}{n_H + n_s} + n_s RT_s \ln \frac{n_s}{n_H + n_s} \tag{A1.3}$$

Assuming $n_H \gg n_0$, which means that the amount of the analyte vapor compared with that of air is trivial, from eqs A1.2 and A1.3, we have,

$$\Delta G_1 = \Delta G_{ii} - \Delta G_i = n_s RT_s \ln \frac{n_s}{n_H} - n_0 RT_s \ln \frac{n_0}{n_H} \tag{A1.4}$$

The entropy change in Step 1 is

$$\Delta S_1 = -\frac{\partial \Delta G_1}{\partial T_s} = -n_s R \ln \frac{n_s}{n_H} + n_0 R \ln \frac{n_0}{n_H} \tag{A1.5}$$

ΔS_2, Entropy Change in Step (2). Step (2) is the cooling of the analyte vapor (n_f) from temperature T_s to T_f as shown in Figure A1.1 (2). The entropy change in this step is the sum of the entropy change of the analyte (ΔS^{sys}) and that of the surrounding environment (ΔS^{surr}),

$$\Delta S^{sys} = \int_{T_s}^{T_f} \frac{dH}{T} = n_f \int_{T_s}^{T_f} C_p \frac{dT}{T} = n_f C_p \ln \frac{T_f}{T_s} \tag{A1.6}$$

$$\Delta S^{surr} = \frac{Q}{T_f} = n_f C_p \left(\frac{T_s - T_f}{T_f} \right) \tag{A1.7}$$

From eqs A1.6 and A1.7, we have,

$$\Delta S_2 = \Delta S^{sys} + \Delta S^{surr} = n_f C_p \left(\frac{T_s - T_f}{T_f} + \ln \frac{T_f}{T_s} \right) \tag{A1.8}$$

where C_p is the heat capacity of the analyte. For simplicity, we assume that C_p is constant from T_f to T_s.

ΔS_3, Entropy Change in Step (3). Step (3) is the absorption of the analyte (n_f) by a liquid coating at constant temperature (T_f) and pressure (P_s), shown in Figure A1.1 (3). The change of Gibbs free energy is,

$$\Delta G_3 = n_f RT_f \ln \frac{p_f}{P_s} \tag{A1.9}$$

and the entropy change in step 3 is

$$\Delta S_3 = -\frac{\partial \Delta G_3}{\partial T_f} = -n_f R \ln \frac{p_f}{P_s} \tag{A1.10}$$

where the p_f is the vapor pressure of the analyte in equilibrium with the analyte dissolved in the coating and is determined by Henry's law,

$$p_f = K_f C_f = K_f \frac{n_f}{V_f} \tag{A1.11}$$

where C_f is the analyte concentration in the coating; V_f is the volume of the coating; K_f is the Henry's constant for the analyte in the coating at the temperature of T_f. K_f can also be expressed as the analyte's gas/coating partition coefficient $K_f = RT_f/K_0$,

$$p_f = \frac{RT_f n_f}{K_0 V_f} \tag{A1.12}$$

Assuming ideal gases in the headspace and the condition $n_H \gg n_0$, we can express the initial pressure, P_s, as:

$$P_s = \frac{n_H R T_s}{V_s} \tag{A1.13}$$

Substituting p_f and P_s from eqs A1.12 and A1.13 into eq A1.10, we can rewrite eq A1.10 as:

$$\Delta S_3 = -n_f R \ln \frac{T_f n_f V_s}{T_s V_f K_0 n_H} \tag{A1.14}$$

A.2 Analysis of Diffusion in and Near the Fiber Coating, without or with "In-Fibre" Derivatization

A solution for the diffusion process through the region in and near to the fibre coating treated as a composite cylinder can be used to model extraction in several situations. Figure A2.1 shows the geometrical configuration of the model, with radii of the different layer boundaries.

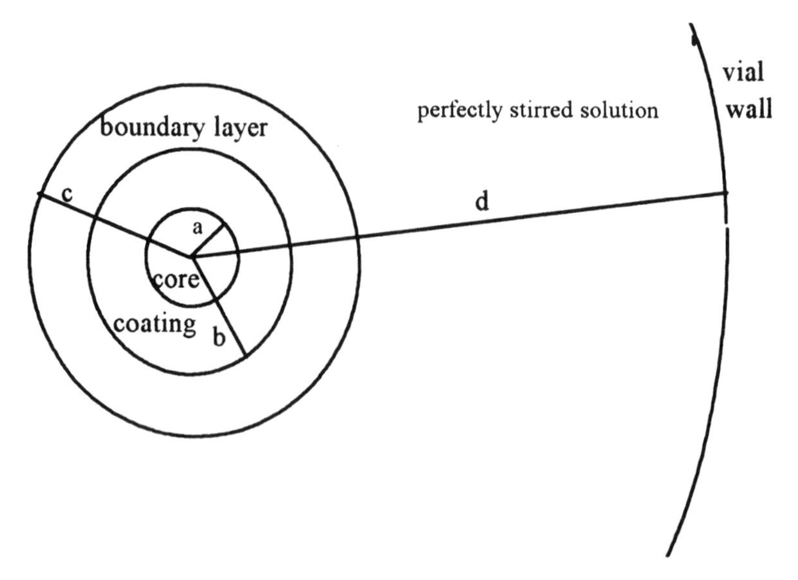

Figure A2.1 The configuration of the boundary value problem for the diffusion process to model the SPME extraction system.

For extraction from a perfectly agitated solution the boundary layer can be of 0 thickness, i.e. c = b. For extraction from a stationary, unstirred solution the outside layer can be the entire sample, i.e. c = d. For extraction from a stirred solution a fibre will have a boundary region where the solution moves slowly near the fibre surface and faster further away until the analyte is practically perfectly mixed in the bulk solution by convection. This boundary region can be modeled as a layer of stationary solution ($b < r < c$) surrounded by perfectly mixed solution ($c < r < d$)[1]. If a compound has high affinity for a fibre's silica core but not the fibre coating, adsorption on the silica core can be modeled as diffusion into a layer, and the fibre coating can be considered a second outside boundary layer. The following mathematical solution also includes the case when a derivatization reaction occurs in the "inside" layer ($a < r < b$) with reaction rate k_1, or in the "outside" layer ($b < r < c$) with reaction rate k_2. To model simple extraction with no derivatization the reaction rates k_1 and k_2 are set to 0.

Since this mathematical solution can be applied to different situations, where the inner and outer layers can represent different media, the parameters use subscripts "1" and "2" instead of the "f", "g", "s" used in the rest of this book.

A.2.1 Statement of the Problem

Following is a modification of a solution published by S. S. Penner and S. Sherman[2]. To be consistent with the notation in the reference, let $u_1(r, t)$ represent the concentration of analyte in the inner layer $a \leq r \leq b$, let $u_2(r, t)$ represent the concentration of analyte in the outer layer $b < r \leq c$. The perfectly stirred region is treated as a cylinder around the fibre of the same height, L, as the fibre. Let $u_3(t)$ the concentration of analyte in the perfectly stirred region $c < r \leq d$.

At time $t = 0$, the analyte is assumed to be at concentration 0 in the fibre coating,

$$u_1(r, 0) = 0, a < r \leq b \tag{A2.1}$$

and at concentration C_0 in the outer layer and bulk solution,

$$u_2(r, 0) = C_0, b < r \leq c \tag{A2.2}$$

Zero adsorption or absorption is assumed for the vial inner surfaces, and the analyte is assumed to be at uniform concentration in the bulk solution,

$$u_3(t) = u_2(c, t) \; t \geq 0 \tag{A2.3}$$

The analyte is assumed to partition into the inner layer with a partition coefficient of K at the inner layer/outer layer interface,

$$D_1 \left.\frac{\partial u_1}{\partial r}\right]_{r=b} = D_2 \left.\frac{\partial u_2}{\partial r}\right]_{r=b}, t > 0 \tag{A2.4}$$

$$u_1(b,t) = K \cdot u_2(b,t), t > 0 \tag{A2.5}$$

The analyte is assumed to diffuse through the outer layer with constant coefficient of diffusion D_2. If the analyte is being derivatized in this "boundary" layer, the reagent concentration is assumed to remain uniform and constant so the reaction rate is proportional to the analyte concentration, hence $\partial u_2(r, t)/\partial t = -k_2 u_2(r, t)$. Therefore the diffusion equation in this layer is:

$$k_2 u_2 + \frac{\partial u_2}{\partial t} = D_2 \left[\frac{\partial^2 u_2}{\partial r^2} + \frac{1}{r}\frac{\partial u_2}{\partial r} \right], b < r < c, t > 0. \tag{A2.6}$$

The analyte is assumed to diffuse into the first layer (usually the fibre coating) with constant coefficient of diffusion D_1. If the analyte is being derivatized in the first layer, the reagent concentration is assumed to remain constant so the reaction rate is proportional to the analyte concentration, hence $\partial u_1(r, t)/\partial t = -k_1 u_1(r, t)$. The diffusion equation in this layer is:

$$k_1 u_1 + \frac{\partial u_1}{\partial t} = D_1 \left[\frac{\partial^2 u_1}{\partial r^2} + \frac{1}{r}\frac{\partial u_1}{\partial r} \right], a < r < b, t > 0. \tag{A2.7}$$

Zero adsorption or absorption at the inner boundary (normally the fibre's silica core) is assumed:

$$\left.\frac{\partial u_1}{\partial r}\right]_{r=a} = 0, t \geq 0. \tag{A2.8}$$

A.2.2 Solution

Following is a modification of a solution published by S. S. Penner and S. Sherman.[2]

First, $v_i e^{-kt}$ is substituted for u_1 and for u_2 to eliminate the "$k_1 u_1$" and "$k_2 u_2$" terms from (A2.6) and (A2.7), then separation of variables leads to the following general solutions of eq.'s (A2.6) and (A2.7):

$$v_1(r,t)e^{-k_1 t} = [A_1 J_0(\lambda_1 r) + B_1 Y_0(\lambda_1 r)]e^{-D_1\lambda_1^2 t - k_1 t}$$

$$v_2(r,t)e^{-k_2 t} = [A_2 J_0(\lambda_2 r) + B_2 Y_0(\lambda_2 r)]e^{-D_2\lambda_2^2 t - k_2 t}$$

where J and Y are the Bessel functions of the first and second kind. A necessary condition for eq.'s (A2.4) and (A2.5) to hold for all $t > 0$ is

$$\lambda_2 = \sqrt{\frac{D_1\lambda_1^2 + (k_1 - k_2)}{D_2}} = f(\lambda_1) \tag{A2.9}$$

The complete solution can be expressed as

$$u(r,t) = \sum_j F_j Z_j(r) e^{-D_1\lambda_j^2 t - k_1 t} \tag{A2.10}$$

where the λ_j represent different non-zero values of λ_1, the F_j are coefficients, and

$$
\begin{aligned}
Z_j(r) &= A_{1j}J_0(\lambda_j r) + B_{1j}Y_0(\lambda_j r) && a \le r \le b,\\
&\quad A_{2j}J_0(f(\lambda_j)r) + B_{2j}Y_0(f(\lambda_j)r) && b < r \le c,\\
&\quad Z_j(c) && c < r \le d.
\end{aligned}
$$

Next the λ_j, F_j, and A_{ij}, B_{ij}, $i = 1, 2$, must be determined.

At the boundary $r = c$ the concentration in the perfectly stirred region will equal the initial amount in that region minus the total flux out of the region, all divided by the volume. This relationship can be expressed:

$$
u_2(c, t) = u_3(t) = C_0 - \frac{2\pi c D_2}{\pi(d^2 - c^2)} \int_0^t \left.\frac{\partial u_2}{\partial r}\right]_{r=c-} dt
$$

$$
= C_0 + \sum_j \frac{2cF_j}{(d^2 - c^2) f(\lambda_j)^2} \left(\left.\frac{\partial Z_j}{\partial r}\right]_{r=c-} - \left.\frac{\partial Z_j}{\partial r}\right]_{r=c-, t=0} \right)
$$

for λ_j not zero. Equality for arbitrary t necessitates

$$
\frac{(d^2 - c^2)}{2c} f(\lambda_j)^2 Z_j(c) = \left.\frac{\partial Z_j}{\partial r}\right]_{r=c} \tag{A2.11}
$$

For non-zero λ_j eq.'s (A2.4), (A2.5), (A2.8) and (A2.11) give 4 equations in 4 unknowns to solve for the A_{ij} and B_{ij} which can be written

$$
\begin{bmatrix}
-J_0(\lambda_j b) & -Y_0(\lambda_j b) & KJ_0(f(\lambda_j)b) & KY_0(f(\lambda_j)b) \\
\lambda_j J_1(\lambda_j b) & \lambda_j Y_1(\lambda_j b) & -\mu f(\lambda_j) J_1(f(\lambda_j)b) & -\mu f(\lambda_j) Y_1(f(\lambda_j)b) \\
J_1(\lambda_j a) & Y_1(\lambda_j a) & 0 & 0 \\
0 & 0 & \Phi f(\lambda_j) J_0(f(\lambda_j)c) + J_1(f(\lambda_j)c) & \Phi f(\lambda_j) Y_0(f(\lambda_j)c) + Y_1(f(\lambda_j)c)
\end{bmatrix}
\begin{bmatrix} A_{1j} \\ B_{1j} \\ A_{2j} \\ B_{2j} \end{bmatrix}
= M \begin{bmatrix} A_{1j} \\ B_{1j} \\ A_{2j} \\ B_{2j} \end{bmatrix} = 0
$$

where $\mu = D_2/D_1$ and $\Phi = (d^2 - c^2)/2c$. A non-trivial solution requires λ_j to be such that $\det(M) = 0$, which gives an equation to determine the λ_j, $j = 1, 2, 3, \ldots$. The expression for u(r, t) is real and bounded for arbitrary t if and only if $D_1\lambda_j^2 + k_1 \geqslant 0$. If $k_1 = 0$ (no derivatization in the inner layer) λ_j can be zero but cannot be imaginary, and for $\lambda_0 = 0$ the conditions imply $B_{10} = B_{20} = 0$, $A_{10} = 1$ and $A_{20} = 1/K$. If $k_1 > 0$ (derivatization in the inner layer)

λ_j cannot be zero but can be imaginary, i.e., $\lambda = iz$, $0 < z < \sqrt{\frac{k_1}{D_1}}$. (When λ or $f(\lambda)$ is imaginary the same equations apply by using the following identities:

$$J_0(iz) = I_0(z), J_1(iz) = iI_1(z), Y_0(iz) = iI_0(z) - \frac{2}{\pi}K_0(z), Y_1(iz)$$

$$= -I_1(z) + \frac{2j}{\pi}K_1(z),$$

where I and K are the Modified Bessel functions of the first and second kind.) Now det(M) = 0 implies rank(M) < 4 which implies three of the A's and B's will depend on the fourth so to normalize set $A_{1j} = 1$ for all j.

Using the following integrals from Penner and Sherman's paper[2]

$$\int rZ_jZ_k\,dr = \frac{r}{\lambda_j^2 - \lambda_k^2}\left[Z_j\frac{\partial}{\partial r}Z_k - Z_k\frac{\partial}{\partial r}Z_j\right] \quad \text{and}$$

$$\int rZ_j^2\,dr = \frac{r^2}{2\lambda_j^2}\left(\frac{\partial}{\partial r}Z_j\right)^2 + \frac{r^2}{2}Z_j^2$$

in which we substitute $\lambda_j = \lambda_{1j}$ for a < r < b and $\lambda_j = f(\lambda_{1j})$ for b < r < d, along with the boundary conditions gives

$$\left[\int_a^b + K\int_b^d\right] rZ_jZ_k = 0 \qquad \text{for} \qquad j \neq k$$

and

$$2\left[\int_a^b + K\int_b^d\right] rZ_j^2 = b^2\left(\frac{1}{\lambda_j^2} - \frac{K}{\mu^2 f(\lambda_j)^2}\right) Z_j'(b)^2 - a^2Z_j(a)^2$$

$$+ \left(1 - \frac{1}{k}\right) b^2Z_j(b)^2 + K\left[\frac{f(\lambda_j)^2(d^2 - c^2)^2}{4} + d^2\right] Z_j(c)^2$$

where as before λ_j represents different non-zero values of λ_1. To determine F_j when λ_j is not zero let t = 0, multiply expression (A2.10) by rZ_j and take the integral $\int_a^b + K\int_b^d$ of both sides to isolate

$$F_j = \frac{1}{2\left[\int_a^b + K\int_b^d\right] rZ_j^2} \frac{2C_0KbD_1\lambda_j[J_1(\lambda_j b) + B_{1j}Y_1(\lambda_j b)]}{f(\lambda_j)^2D_2}.$$

To determine F_0 when λ_0 is zero (which is possible only when $k_1 = 0$) let t = 0, multiply expression (A2.10) by r and take the integral $\int_a^d$ of both sides of the equation to isolate

$$F_0 = \frac{KC_0(d^2 - b^2)}{d^2 - b^2 + K(b^2 - a^2)} = \frac{KC_0}{1 + K\frac{V_f}{V_s}}$$

where V_f is the volume of the fibre inner layer shown in Figure A2.1, and V_s is the "equivalent volume" of the perfectly stirred region plus outer boundary layer. This formula completes the solution for u(r, t).

The amount of analyte in the fibre at time t will be $2\pi L \int_a^b u(r, t)r\, dr$, and the total amount of analyte reacted up to time t = T will be $2\pi Lk \int_0^T \int_a^b u(r, t)r\, dr\, dt$.

Note the perfectly stirred region is treated as a cylinder around the fibre of the same height, L, as the fibre. In reality a sample will be around, below and perhaps above the fibre. Since the sample outside the boundary layer is treated as perfectly stirred, however, the exact geometry of the bulk sample does not affect the extraction rate calculated from this model. Therefore in the model the radius of the cylinder around the fibre of the same height, L, as the fibre should be set such that the "equivalent" sample volume in this region has the same capacity as the actual sample. Therefore for the fibre immersed in a volume V_s of water "d" should be such that $\pi L(d^2 - c^2) = V_s - \pi L(c^2 - b^2)$.

A.2.3 Analysis of Diffusion in the Fiber Coating with "In-Fibre" Derivatization

When extracting from air or from the headspace of a well stirred solution the situation corresponds to a boundary layer of thickness 0, that is $c - b = 0$. In these cases the SPME process often can be modeled with a single layer, as shown in Figure A.2.2. The theoretical prediction of the diffusion process for this configuration can be calculated using the result given above in section A.2.2 with c = b. The expressions in A.2.2 with c = b give numerically correct answers, but the expressions are in a complicated form. Repeating the same algorithm to derive the theoretical solution for the system with a single layer produces simpler formulae. This derivation was done, but only the result is presented because the essential mathematical techniques used are already given in A.2.2.

The notation used in this section is similar to the notation in section A.2.2. The perfectly mixed solution is in the region $b < r < d$, and at time $t = 0$ the analyte is assumed to be at concentration C_0 in the mixed solution. Let K be the analyte's distribution constant between coating and sample, and D the analyte's diffusion coefficient in the coating, both assumed constant. If the analyte is being derivatized in the fibre coating, the reagent concentration is assumed to remain constant so the reaction rate is proportional only to the analyte concentration with reaction rate constant "k", hence $\partial u_2(r, t)/\partial t = -k_2 u_2(r, t)$.

When $k = 0$ (no derivatization), if the SPME extraction has negligible effect on the analyte concentration in the sample, the diffusion process into a cylindrical SPME fibre is mathematically the same as the problem which is solved in Carslaw and Jaeger[3], section 13.4.

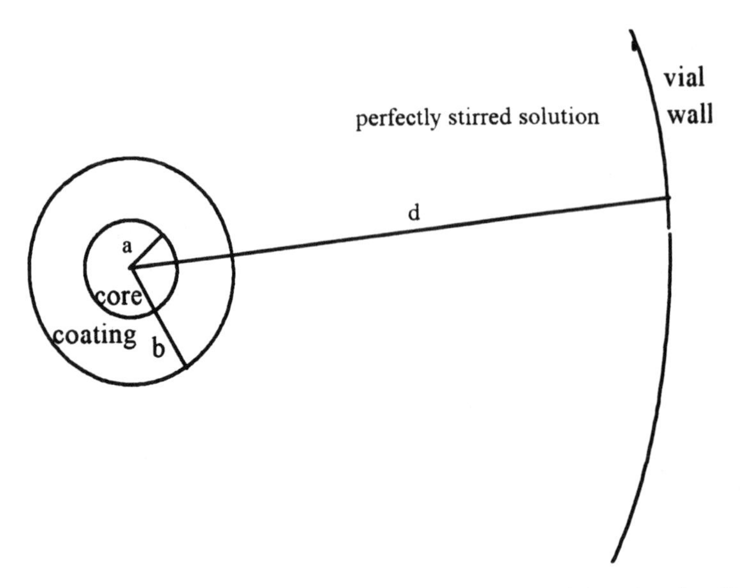

Figure A2.2 The configuration of the boundary value problem for the diffusion process of SPME extraction from a perfectly stirred sample.

When k > 0, the amount of analyte reacted up to time t, R(t), can be expressed with the dimensionless parameters $h = \frac{a^2k}{D}$, $T = \frac{Dt}{a^2}$, $H = \frac{d^2 - b^2}{Ka^2}$ and $\theta = \frac{b}{a}$ as

$$\frac{R(T)}{C_0V_s} = \tag{A2.12}$$

$$1 + 2h\sum_j \frac{e^{-(\alpha_j^2+h)T}}{2\alpha_j 2\left(\frac{\theta^2}{H} + 1\right) + \frac{H}{2}(\alpha_j^2 + h)^2 + \frac{4}{\pi^2\alpha_j\theta}(\alpha_j^2 + h)G_1(\alpha_j)^{-1}G_2(\alpha_j)^{-1}}$$

where
$G_1(\alpha_j) = Y_1(\alpha_j)J_0(\alpha_j\theta) - Y_0(\alpha_j\theta)J_1(\alpha_j)$ and $G_2(\alpha_j) = Y_1(\alpha_j)J_1(\alpha_j\theta) - Y_1(\alpha_j\theta)J_1(\alpha_j)$, and the α_j, j = 1, 2, 3 . . . are the positive real or positive imaginary roots of

$$H(\alpha_j^2 + h)G_1(\alpha_j) + 2\alpha_j\theta G_2(\alpha_j) \tag{A2.13}$$

A.2.4 Approximation to the Result in A.2.3

If derivatization is applied to improve sensitivity then probably $KV_f \ll V_s$. The assumption $V_s/KV_f > 10$ is made throughout this analysis.

The first term of the series for R(T) given in A.2.3 corresponds to the imaginary root of (13), which is the positive root of

$$\frac{2\theta\alpha}{H(h-\alpha^2)} - \frac{I_0(\alpha\theta)K_1(\alpha) + I_1(\alpha)K_0(\alpha\theta)}{I_1(\alpha\theta)K_1(\alpha) - I_1(\alpha)K_1(\alpha\theta)}. \qquad \text{(A2.14)}$$

Let $f(x, \theta) = \dfrac{I_0(x\theta)K_1(x) + I_1(x)K_0(x\theta)}{I_1(x\theta)K_1(x) - I_1(x)K_1(x\theta)}$. $f(\alpha, \theta)$ decreases with α approaching 1 as $x \to \infty$, and $2\theta\alpha/H(h - \alpha^2)$ increases with α to infinity for $0 < \alpha^2 < h$, therefore equation (A2.14) has exactly one positive root.

Consider $f(x, \theta) \cdot x$. This function increases with x and approaches $2\theta/(\theta^2 - 1)$ as $x \to 0$. At $x = s/(\theta - 1)$, where s is a constant, the function $f(x, \theta) \cdot x/[2\theta/\theta^2 - 1)] = f(s/(\theta - 1), \theta) \cdot s(\theta + 1)/(2\theta)$ decreases with θ and approaches $s \cdot \cosh(s)/\sinh(s)$ as $\theta \to 1$. (These results can be derived from the asymptotic expressions for Bessel functions.) Thus $0 < x < s/(\theta - 1) \Rightarrow 2\theta/(\theta^2 - 1) < f(x, \theta) \cdot x < s \cdot \cosh(s)/\sinh(s) \cdot 2\theta/(\theta^2 - 1)$ for all $\theta > 1$.

Now consider $f(x, \theta)$. $f(x, \theta)$ decreases with x, approaches 1 as $x \to \infty$, decreases with θ and approaches 1 as $\theta \to \infty$. At $x = s/(\theta - 1)$, $f(x, \theta) = f(s/(\theta - 1), \theta)$ increases with θ, approaches $\cosh(s)/\sinh(s)$ as $\theta \to 1$ and approaches $I_0(s)/I_1(s)$ as $\theta \to \infty$. Thus $x > s/(\theta - 1) \Rightarrow 1 < f(x, \theta) < I_0(s)/I_1(s)$ for all $\theta > 1$.

To determine the criteria for $\alpha_1 < s/(b - a)$ or $\alpha_1 > s/(b - a)$, solving for α in the equation (A2.14) will show

$$h(\theta - 1)^2 < s^2 + \frac{s \tanh(s)}{H(\theta^2 - 1)} \Rightarrow f(x, \theta) \cdot x < s \cdot \coth(s) \cdot 2\theta/(\theta^2 - 1)$$

$$\Rightarrow \alpha_1 < s/(\theta - 1),$$

and

$$h(\theta - 1)^2 > s^2 + s\frac{2\theta(\theta - 1)}{H} \Rightarrow f(x, \theta) < I_0(s)/I_1(s) \qquad \text{(A2.15)}$$

$$\Rightarrow \alpha_1 > s/(\theta - 1)$$

To determine the time for the amount of analyte reacted to reach 95% consider two cases:

Case 1 $\alpha_1 < s/(\theta - 1)$

The limits determined for $f(x, \theta)x$ give upper and lower limits for α_1. Solving for α in the equation (A2.14) gives $\alpha_1 = \sqrt{\dfrac{h}{\dfrac{\theta^2 - 1}{H\Phi} + 1}}$, where Φ is between 1 and $s \cdot \coth(s)$, hence the first term reaches 95% at

$$T = \frac{-\ln(0.05)}{h}\left(\frac{H\Phi}{\theta^2 - 1} + 1\right).$$

The series expression for the amount reacted in the result given in A2.3 above gives 0 at T = 0. Therefore the series at T = 0 shows the sum of all coefficients of exponentials in T must be −1. The expression derived for R(T) before converting to dimensionless parameters was

$$\frac{R(T)}{C_0V_s} = 1 + \frac{k(d^2 - b^2)}{KD}\sum_{\lambda}\frac{z(b)^2}{\lambda^2\left[\int_a^b + K\int_b^d\right]rZ(r)^2}e^{-(k+D\lambda^2)t}.$$

This expression shows that at the real roots of (A2.13) all coefficients of the $e^{-(h+\alpha\cdot\alpha)T}$ terms are positive. Since at T = 0 the value of R(T) is 0, the coefficient of the $e^{-(h+\alpha\cdot\alpha)T}$ term at the imaginary root of (A2.13) must be negative. Another conclusion is that the sum of all the positive coefficients cannot be greater than the one negative coefficient.

For all positive coefficients of the $e^{-(h+\alpha\cdot\alpha)T}$ terms corresponding to real positive values of α the exponential can be at most e^{-hT}. For the first term in the series the exponential is found by substituting for α_1 according to the expression given above. Also, assume $KV_f < V_s/10$, that is $(\theta^2 - 1)/H < 0.1$. The result is the maximum exponential value for the first term is $e^{-hT/11}$. Therefore when the first term exponential reaches 0.05, the subsequent exponential terms with positive coefficients can be at most 0.05^{11}, vanishingly small. Since for all values of T the total amount reacted can never be negative, the minimum the negative coefficient can be is greater than the maximum value of $-1/(f - f^{11})$ where f is between 0 and 1, which can be shown to be −1.4. Also, since R(T) is 0 at T = 0 the negative coefficient must be at most −1. Therefore the value of T at which the entire series expression for R(T) reaches 95% of its final equilibrium value is between

$$T = \frac{-\ln(0.05)}{h}\left(\frac{H\Phi}{\theta^2 - 1} + 1\right) \quad \text{and} \quad T = \frac{-\ln(0.05/1.4)}{h}\left(\frac{H\Phi}{\theta^2 - 1} + 1\right).$$

Using the limits of $(\theta^2 - 1)/H < 0.1$ and the limits on Φ given above, the time to reach 95% amount reacted is

$$t_e = t_{95\%} = \Omega\frac{V_s}{kKV_f}$$

where Ω is between 3 and 3.7s · coth(s).

Case 2 $\alpha_1 > s/(b - a)$

The limits determined for $f(x, \theta)$ give upper and lower limits for α_1. Solving for α in the equation (A2.14) gives $\alpha_1 = \frac{\theta}{H\Theta}\left(\sqrt{1 + h\left(\frac{H}{\theta}\Theta\right)^2} - 1\right)$, hence the value of T for the first term to reach 5% of its initial value is $T = \frac{-\ln(0.05)}{2h}\left(1 + \sqrt{1 + h\left(\frac{H}{\theta}\Theta\right)^2}\right)$, where Θ is between 1 and $I_0(s)/I_1(s)$.

As derived for Case I, the value of T at which the entire series expression for the amount reacted reaches 95% is between the expression just given and the same with ln(.05) replaced by ln(.05/1.4). The condition specified in (A2.15) along with the assumption $(\theta^2 - 1)/H < 0.1$ implies $h^{1/2}H/\theta > 10s$, so if s > 0.24 the result simplifies to $t_e = t_{95\%} = \Psi \dfrac{V_s}{\sqrt{kD}\,KA_f}$ where Ψ is between 3 and $3.7I_0(s)/I_1(s)$.

Summary

The first approximation formula is more accurate for small s, the second for large s, and they are equally accurate for s = 1.51. At s = 1.51, if $\dfrac{KV_f}{V_s} < 0.1$, the time for the amount of analyte reacted to reach 95% of total initial amount is:

$$\text{If } \frac{k}{D}(b-a)^2 < 2.3 + 1.4\frac{KV_f}{V_s} \quad \text{then } t_e = t_{95\%} = 4.6\frac{V_s}{kKV_f} \pm 35\%$$

$$\text{If } \frac{k}{D}(b-a)^2 > 2.3 + 1.5\frac{KA_f(b-a)}{V_s}$$

$$\text{then} \quad t_e = t_{95\%} = 4.6\frac{V_s}{\sqrt{kD}\,KA_f} \pm 35\%.$$

A.3 One-Dimensional Analysis of Diffusion in a Headspace SPME System

This analysis treats the fibre coating as a flat membrane at one end of the vial. The vial is symmetric about its axis, vial surfaces are assumed to have no adsorption or absorption of the analyte, and any radial diffusion is considered instantaneous, so the diffusion system is 1 − dimensional. The configuration for the 1 dimensional model is shown in Figure A3.1.

A.3.1 Statement of the Problem

The sample is treated as stationary. Let $C_s(x, t)$ be the analyte's concentration in the sample, D_s the analyte's diffusion coefficient in the sample, and L_s the sample depth. Initial concentration in the sample is C_{s0}, i.e. $C_s(x, 0) = C_{s0}$ for $-L_s < x < 0$. The vial bottom is assumed to have no interactions with the analyte and no flux, thus at the boundary $x = -L_s$ the condition is $\left.\dfrac{\partial}{\partial x} C_s(x,t)\right|_{x=L_s} = 0$, for t > 0.

Headspace is treated as stationary. Let $C_g(x, t)$ be the analyte's concentration in the headspace, D_g the analyte's diffusion coefficient in the headspace,

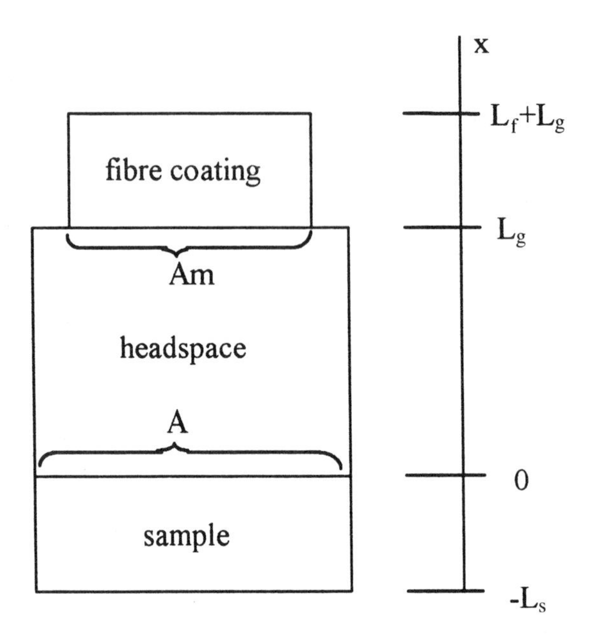

Figure A3.1 The configuration of the simplified 1-dimensional model for headspace SPME.

and L_g the headspace length. Headspace has the same cross-sectional area as the sample. Initial concentration in headspace is C_{g0}, i.e. $C_g(x, 0) = C_{g0}$ for $0 < x < L_g$. At the sample-headspace interface, the boundary $x = 0$, the conditions are $D_s \frac{\partial}{\partial x} C_s(x, t) \Big|_{x=0} = D_g \frac{\partial}{\partial x} C_g(x, t) \Big|_{x=0}$ and $C_g(0, t) = K_{gs}C_s(0, t)$ where K_{gs} is the gas to water distribution constant (Henry's constant) of the analyte.

The coating phase is treated as stationary. Let F be the ratio of membrane surface area, A_m, to vial cross-sectional area, A, such that $F = A_m/A$. The term F is introduced only to account for the small volume of the fibre coating relative to the sample and headspace volumes, radial diffusion is not considered. Note the model presented in Figure 3.17 corresponds to a value of F = 1, and all calculations and results presented in Chapter 3 based on this model use the value F = 1.

Let $C_f(x, t)$ be the analyte's concentration in the membrane, D_f the analyte's diffusion coefficient in the membrane, and L_f the membrane thickness. Initial concentration in the membrane is assumed 0, i.e. $C_f(x, 0) = 0$ for $L_g < x < L_g + L_f$. At the headspace-membrane interface, the boundary $x = L_g$, the conditions are $D_s \frac{\partial}{\partial x} C_s(x, t) \Big|_{x=L_g} = FD_f \frac{\partial}{\partial x} C_f(x, t) \Big|_{x=L_g}$ and $C_f(L_g, t) = K_{fg}C_g(L_g, t)$ where K_{fg} is the fibre to gas distribution constant of the analyte. The outside surface of the membrane is assumed to have no interactions with the analyte and no flux, thus at the boundary x = $L_g + L_f$ the condition is

$$\frac{\partial}{\partial x} C_f(x,t)\bigg]_{x=L_g+L_f} = 0 \text{ for } t > 0.$$

The diffusion equations in the three phases are $D_i \frac{\partial^2}{\partial x^2} C_i(x,t) = \frac{\partial}{\partial t} C_i(x,t)$, where "i" is "s", "g" or "f". Note the analyte is assumed to have equilibrated between the sample and the headspace before the start of SPME extraction. Let C_0 be the original concentration of the analyte in the sample, before it was transferred to the vial. At the start of SPME, the initial concentration of analyte in the sample would be $C_{s0} = \frac{L_s C_0}{K_{gs} L_g + L_s}$ and the initial concentration in headspace would be $C_{g0} = K_{gs} C_{s0}$.

Also, the one-dimensional model has parameters of length L_s, L_g, and FL_f. These lengths are directly proportional to the volumes of sample, headspace and fibre, respectively, since the vial is symmetric about its axis. Therefore in all results given below, the numerical values of V_s, V_g and V_f can be substituted for L_s, L_g, and FL_f, respectively, so long as these substitutions are made everywhere.

A3.2 Solution

This solution follows the method outlined in Carslaw and Jaeger.[4]

Apply the Laplace transformation to all equations, and let c_1, c_2 and c_3 be the transforms of C_s, C_g and C_f. To maintain consistent notation define $C_{10} = C_{s0}$, $C_{20} = C_{g0}$. The equations and the boundary conditions to satisfy are:

$$\frac{\partial^2 c_i}{\partial x^2} - q_i^2 c_i = -\frac{C_{i0}}{C_i} \quad \text{for } i = 1,2 \tag{A3.1}$$

$$\frac{\partial^2 c_i}{\partial x^2} - q_i^2 c_i = 0 \quad \text{for } i = 3 \tag{A3.2}$$

where $q_i = \sqrt{\frac{p}{D_i}}$, for i = 1, 2, 3,

$$\frac{\partial}{\partial x} c_1 = 0 \quad \text{at } x = -L_s \tag{A3.3}$$

$$K_{gs} c_1 = c_2 \quad \text{and} \quad D_s \frac{\partial}{\partial x} C_1 = D_g \frac{\partial}{\partial x} C_2 \quad \text{at } x = 0 \tag{A3.4}$$

$$K_M c_2 = c_3 \quad \text{and} \quad FD \frac{\partial}{f\,\partial x} C_3 = D_g \frac{\partial}{\partial x} C_2 \quad \text{at } x = L_g \tag{A3.5}$$

$$\frac{\partial}{\partial x} C_3 = 0 \quad \text{at } x = L_g + L_f \tag{A3.6}$$

Note that $p = D_s q_1^2 = D_g q_2^2 = D_f q_3^2$, so that we can write $q_1 = q \cdot \sqrt{D_g/D_s} = q \cdot \Psi$ and $q_3 = q \cdot \sqrt{D_g/D_f} = q \cdot \Phi$, where $q = q_2$, $\Psi = \sqrt{D_g/D_s}$, $\Phi = \sqrt{D_g/D_f}$.

The general solutions of (A3.1) are $c_i(x, p) = A_i \sinh(q_i [x + a_i]) + B_i \cosh(q_i [x + b_i]) + C_{i0}/p$ for $i = 1, 2$, and the general solution of (A3.2) is $c_3(x, p) = A_3 \sinh(q_3 [x + a_3]) + B_3 \cosh(q_3 [x + b_3])$.

To satisfy (A3.3) let $A_1 = 0$ and $b_1 = L_s$, giving $c_1(x, p) = B_1 \cosh(q\Psi[x + L_s]) + C_{10}/p$.

To satisfy (A3.4) let $a_2 = b_2 = 0$ giving $B_2 = \dfrac{K_{gs}\Psi\cosh(q\Psi L_s)}{\sinh(q\Psi L_s)} A_2$ and

$$B_1 = \frac{\Psi}{\sinh(q\Psi L_s)} A_2.$$

To satisfy (A3.6) let $A_3 = 0$ and $a_3 = b_3 = -(L_f + L_g)$.

To satisfy (A3.5) the two equations of (A3.5) are used to solve for A_2 and B_3 giving

$$A_2 = C_{10} F K_{fg} K_{gs} \frac{\sinh(q\Psi L_s)\sinh(q\Phi L_f)}{D_g q^2 \Delta(p)}$$

$$B_3 = C_{10} \Phi K_{fg} K_{gs} \frac{\sinh(q\Psi L_s)\cosh(qL_g) + \Psi K_{gs} \cosh(q\Psi L_s(\sinh(qL_g)}{D_g q^2 \Delta(p)}$$

where

$$\begin{aligned}\Delta(p) = {} & F D_{fg} \sinh(q\Phi L_f)\sinh(qL_g)\sinh(q\Psi L_s) \\ & + \Phi\Psi K_{gs}\cosh(q\Psi L_s)\cosh(q\Phi L_f)\sinh(qL_g) \\ & + \Phi\cosh(q\Phi L_f)\cosh(qL_g)\sinh(q\Psi L_s) \\ & + F K_{fg}\Psi K_{gs}\sinh(q\Phi L_f)\cosh(qL_g)\cosh(q\Psi L_s)\end{aligned}$$

To solve for C_s, C_g and C_f we use the Laplace Inversion theorem. We assume the integrands have only simple poles at zero and at the roots of $\Delta(p)$ along the negative real axis. This assumption is reasonable since the resulting expressions for C_s, C_g and C_f would not appear to be real and bounded if the poles were at complex or positive values of p. No mathematical proof is offered, however, so as with many solutions to boundary value problems the final answers should be checked to ensure they satisfy the boundary conditions. Computation is necessary only for the conditions at $t = 0$. Thus by the Residue theorem the result is:

$$\begin{aligned}C_s(x, t) = {} & \frac{C_0 L_s}{K_{gs} L_g + L_s + K_{fg} K_{gs} F L_f} \\ & + 2C_{10} K_{fg} K_{gs} F \Psi \sum_{z_i} e^{-D_g z_i^2 t} \frac{\sin(z_i \Phi L_f)}{z_i \frac{\partial}{\partial e} \Lambda(z) \rfloor_{z=z_i}} \cos(z_i \Psi[x + L_s])\end{aligned}$$

$$C_s(x, t) = \frac{C_0 L_s K_{gs}}{K_{gs} L_g + L_s + K_{fg} K_{gs} F L_f}$$

$$+ 2C_{10} K_{fg} K_{gs} F \Psi \sum_{z_i} e^{-D_g z_i^2 t} \frac{\sin(z_i \Phi L_f)}{z_i \frac{\partial}{\partial e} \Lambda(z) \rfloor_{z=z_i}}$$

$$[-\sin(z_i \Psi L_s)\sin(z_i x) + \Psi K_{gs} \cos(z_i \Psi L_s)\cos(z_i x)]$$

and

$$C_f(x, t) = \frac{C_0 L_s K_{gs} K_{fg}}{K_{gs} L_g + L_s + K_{fg} K_{gs} F L_f}$$

$$+ 2C_{10} K_{fg} K_{gs} \Phi \sum_{z_i} e^{-D_g z_i^2 t}$$

$$\frac{[\sin(z_i \Phi L_s)\cos(z_i L_g) + \Psi K_{gs} \cos(z_i \Psi L_s)\sin(z_i L_g)]}{z_i \frac{\partial}{\partial z} \Lambda(z) \rfloor_{z=z_i}}$$

$$\cos(z_i \Phi[x - L_g - L_f])$$

where the z_i are the positive roots of

$$\Lambda(z) = -F K_{fg} \sin(z \Phi L_f)\sin(z L_g)\sin(z \Psi L_s)$$
$$+ \Phi \Psi K_{gs} \cos(z \Psi L_s)\cos(z \Phi L_f)\sin(z L_g)$$
$$+ \Phi \cos(z \Phi L_f)\cos(z L_g)\sin(z \Psi L_s)$$
$$+ F K_{fg} \Psi K_{gs} \sin(z \Phi L_f)\cos(z L_g)\cos(z \Psi L_s)$$

From the above solution the amount extracted at time t can be calculated by

$$\text{Amount}(t) = A_m \int_{L_g}^{L_g + L_f} C_f(x, t) dx$$

A.3.3 Approximation by Assuming a Constant Concentration Gradient in the Headspace

A simple formulae is derived to predict approximately the equilibration time of headspace SPME.

To simplify the mathematics, analyte concentration in the headspace is assumed to follow a straight line from $x = 0$ to $x = L_g$ and the capacity of the headspace is assumed to have no effect on the extraction rate. (This approximation overestimates the diffusion time through the headspace, but the error should be negligible when the headspace contains only a small percentage of the amount extracted.) Diffusion time in the membrane is assumed negligible, therefore the membrane is treated as a perfectly stirred phase. Let $C_f(t)$ be the analyte concentration in the coating.

With these simplifications the boundary value problem stated in A.3.1 becomes simpler. This newly formulated boundary value problem is mathematically the same as a problem which was solved by J. C. Jaeger.[5] From his solution, the concentration in the fibre coating is given by

$$\frac{C_f(t)}{C_{s0}K_{fg}K_{gs}} = 1 - \frac{1}{1+k} - 2kH^2\sum_{\alpha_i}\frac{e^{-\alpha^2 T}}{P_s}$$

where $T = D_s t/L_s^2$, $k = L_s/(L_f F K_{gs} K_{fg})$, $H = L_s K_{gs} D_g / L_g D_s$, $P_s = \alpha^4 + (H + H^2 - 2kH)\alpha^2 + kH^2(k+1)$, and $\alpha_1, \alpha_2, \alpha_3 \ldots$ are the positive roots of $\tan \alpha = \frac{H\alpha}{\alpha^2 - kH}$. The value of T at which fibre concentration reaches 95% of equilibrium, $T_{95\%}$, is a function of parameters k and H only. Computer calculation of the plot "$T_{95\%}$ versus k, H" showed

$$T_{95\%} = \left(\frac{3}{H} + \frac{1.5}{k/100 + 1}\right)\frac{1}{1+k} \pm 30\%$$

for all values of H, k > 0. Expressed in the original parameters of the headspace SPME model the result is

$$t_{95\%} = 3\left(\frac{L_g}{K_{gs}D_g} + \frac{1}{2D_s}\frac{K_{fs}FL_F L_s}{0.01L_s + K_{fs}FL_f}\right)\frac{K_{fs}FL_f L_s}{L_s + K_{fs}FL_f} \pm 30\%$$

Further computation determined if $L_s < 170K_{fs}FL_f$ a simpler formula is

$$t_{95\%} = 3\left(\frac{L_g}{K_{gs}D_g} + \frac{L_s}{3.3D_s}\right)\frac{K_{fs}FL_f L_s}{L_s + K_{fs}FL_f} \pm 50\%,$$

and if $K_{fs}FL_f < L_s < 20K_{fs}FL_f$ then $t_{95\%} = 1.8\left(\frac{L_g}{K_{gs}D_g} + \frac{L_s}{1.6D_s}\right)K_{fs}FL_f \pm 50\%$.

If the parameters satisfy $H < 0.2$ then $t_{95\%} = 3\left(\frac{L_g}{K_{gs}D_g}\right)\frac{K_{fs}F_{fs}L_f L_s}{L_s + K_{fs}FL_f} \pm 40\%$.

This result implies that if $H < 0.2$ then diffusion time in the sample has no significant effect on the extraction rate.

A.3.4 Approximation by Treating the Sample as Perfectly Stirred

As noted in the preceeding paragraph, under certain conditions diffusion in the sample does not affect the extraction rate in headspace SPME.

The same basic model configuration and parameters as explained above in section A.3.1 are assumed, except sample is assumed perfectly stirred and of infinite volume, so that it maintains constant concentration C_{s0}. Diffusion time in the membrane is assumed negligible, so the analyte concentration is assumed to be uniform throughout the coating.

This same problem is solved in the Carslaw and Jaeger[6], section 3.13, eq. (8). The concentration in the fibre is given by

$$C_f(t) = K_{fg}C_{20}\left[1 - 2h\sum_{n=1}^{\infty}\frac{e^{-\alpha_n^2 T}}{\alpha_n^2 + h^2 + h}\right]$$

where $T = tD_g/L_g^2$, $h = L_g/(FL_fK_{fg})$, and α_n, n = 1, 2, 3, . . . are the positive roots of $\alpha \tan(\alpha) = h$.

The value of T at which fibre concentration reaches 95% of equilibrium, $T_{95\%}$, is a function of parameter h only. Computer calculation of the plot "$T_{95\%}$ versus h" shows $T_{95\%} = 2.9/h + 0.29 \pm 19\%$ for h from 0 to 15.

The equilibration time for perfectly stirred sample and static headspace will approach a maximum as h goes to infinity. This limit corresponds to extraction from an infinite, purely gaseous sample. This problem is also solved in Carslaw and Jaeger.[7] The concentration in the fibre is given by

$$C_f(t) = K_{fg}C_{20}[1 - e^T erfc(\sqrt{T})]$$

where $T = D_g t/(K_{fg}L_fF)^2$. A simple calculation shows the equilibration time for this system is $t_e = t_{95\%} = 126(K_{fg}L_fF)^2/D_g$.

The formula $T_{95\%} = 2.9/h + 0.29$ for the equilibration time will reach $126(K_{fg}FL_f)^2/D_g$, known to be the maximum possible as explained above, when h reaches 15. Therefore, the equilibration time for extraction from a perfectly stirred sample and static headspace, neglecting diffusion time in the coating, is approximately

$$t_e = t_{95\%} = \begin{Bmatrix} \dfrac{2.9}{D_g}\left(K_{fg}FL_fL_g + \dfrac{L_g^2}{10}\right) \pm 19\% & L_g < 15K_{fg}FL_f \\ \dfrac{126}{D_g}(K_{fg}FL_f)^2 & L_g > 15K_{fg}FL_f \end{Bmatrix}.$$

The above formula also shows if $L_g > 12K_{fg}FL_f$ then $t_e = 100(K_{fg}FL_f)^2/D_g \pm 25\%$, which is the formula 3.53 given in the book for F = 1.

References

1. McGraw-Hill eds. *Encyclopedia of Science and Technology.* (McGraw-Hill: New York, Toronto, 1992).

2. S. S. Penner, S. Sherman, *Journal of Chemical Physics* **15** 569 (1947).

3. H. S. Carslaw, J. C. Jaeger, *Conduction of Heat in Solids, 2nd Edition* (Clarion Press: Oxford, 1959), section 13.4, p. 332.

4. H. S. Carslaw, J. C. Jaeger, *Conduction of Heat in Solids, 2nd Edition* (Clarion Press: Oxford, 1959), section 12.8, p. 319.

5. J. C. Jaeger, *Proc. Camb. Phil. Soc.* **45** 43 (1945).

6. H. S. Carslaw, J. C. Jaeger, *Conduction of Heat in Solids, 2nd Edition* (Clarion Press: Oxford, 1959), section 3.13, eq. (8), p. 128.

7. H. S. Carslaw, J. C. Jaeger, *Conduction of Heat in Solids, 2nd Edition* (Clarion Press: Oxford, 1959), section 12.4, p. 306.

APPENDIX

B

List of Articles Published on Applications of SPME

The following bibliography is organized according to subject matter. Only articles from scientific journals are included, no conference proceedings or commercial publications are listed. Within each subsection articles are listed in chronological order by date of publication, beginning with the earlies. Some articles appear under more than one heading.

Environmental Analysis

Air Analysis

1. C. L. Arthur, L. Killam, K. Buchholz, D. Potter, M. Chai, Z. Zhang and J. Pawliszyn, "Solid Phase Microextraction: An Attractive Alternative, or Applications of Automated SPME to Environmental Analysis" *Amer. Environ. Lab.* **11,** 10 (1992).
2. C. L. Arthur, M. Chai and J. Pawliszyn, "Solventless Injection Techniques for Microcolumn Separations" *J. Microcolumn Sep.* **5,** 51 (1993).
3. M. Chai, C. Arthur, J. Pawliszyn, R. Belardi and K. Pratt, "Determination of Volatile Chlorinated Hydrocarbons in Air and Water with Solid-Phase Microextraction" *Analyst* **118,** A1501 (1993).
4. M. Chai and J. Pawliszyn, "Analysis of Environmental Air Samples by SPME and GC-ITMS" *Environ. Sci. Technol.* **29,** A693 (1995).
5. L. Pan, M. A. Adams and J. Pawliszyn, "Determination of Fatty Acids using SPME" *Anal. Chem.* **67,** 4396 (1995).

6. P. Martos and J. Pawliszyn, "Calibration of SPME for Air Analyses Based on Physical Chemical Properties of the Coating" *Anal. Chem.*, **69,** 206 (1997).
7. J. Czerwinski, B. Zygmunt and J. Namiesnik, "Application of solid phase microextraction for determination of volatile halogenated hydrocarbons in air and water of an indoor swimming pool" *Fresenius Environ. Bull.* **5,** 55 (1996).
8. J. P. Snell, W. Frech and Y. Thomassen, "Performance improvements in the determination of mercury species in natural gas condensate using an online amalgamation trap or solid-phase micro-extraction with capillary gas chromatography-microwave-induced plasma atomic emission spectrometry" *Analyst* **121,** 1055 (1996).
9. T. J. Clark and J. E. Bunch, "Quantitative determination of phenols in mainstream smoke with solid-phase microextraction-gas chromatography-selected ion monitoring mass spectrometry" *J. Chromatogr. Sci.* **34,** 272 (1996).
10. C. Malosse, P. Ramirez-Lucas, D. Rochat and J. P. Morin, "Solid-phase microextraction, an alternative method for the study of airborne insect pheromones (Metamasius hemipterus, Coleoptera, Curculionidae)" *J. High Resolut. Chromatogr.* **18,** 669 (1995).

Soil Analysis

1. Z. Zhang and J. Pawliszyn, "Headspace Solid Phase Microextraction" *Anal. Chem.* **65,** 1843 (1993).
2. Z. Zhang and J. Pawliszyn, "Analysis of Organic Compounds in Environmental Samples Using Headspace Solid Phase Microextraction" *J. High Resolut. Chromatogr.* **16,** C689 (1993).
3. P. Popp, K. Karsten and O. Gudrun, "Application of SPME and GC with EC and MS detection for the determination of hexachlorocyclohexanes in soil solutions" *J. Chromatogr., A* **687,** 133 (1994).
4. K. James and M. A. Stack, "The determination of volatile organic compounds in soils using solid phase microextraction with gas chromatography-mass spectrometry" *J. High Resolut. Chromatogr.* **19,** 515 (1996).
5. A. Fromberg and J. O. Madsen, "Analysis of chlor- and nitroanilines and benzenes in soils by headspace solid phase microextraction" *J. Chromatogr., A* **746,** 71 (1996).
6. J. D. Gaynor, D. A. Cancilla, G. R. B. Webster, L. P. Sarna, K. N. Graham, H. Y. F. Ng, C. S. Tan and C. F. Drury, "Comparative solid phase extraction, solid phase microextraction, and immunoassay analyses of Metolachlor in surface runoff and tile drainage" *J. Agric. Food Chem.* **44,** 2243 (1996).
7. S. Bengtsson, T. Bergloef and T. Sjoeqvist, "Predicting the leachability of pesticides from soils using near-infrared reflectance" *J. Agric. Food Chem.* **44,** 2260 (1996).

8. H. Honda, "Application of SPME to examining volatile organic compounds in landfills" *Aichi-ken Kogai Chosa Senta Shoho* **23,** 57 (1995).

Water Analysis

1. R. P. Belardi and J. Pawliszyn, "The Application of Chemically Modified Fused Silica Fibres in the Extraction of Organics from Water Matrix Samples and their Rapid Transfer to Capillary Columns" *Water Pollut. Res. J. Can.* **24,** 179 (1989).
2. C. L. Arthur and J. Pawliszyn, "Solid Phase Microextraction with Thermal Desorption Using Fused Silica Optical Fibers" *Anal. Chem.* **62,** 2145 (1990).
3. J. Pawliszyn, *Instrumentation for Trace Organic Monitoring,* R. Clement, M. Siu, H. Hill eds. (Lewis Publishers, Boca Raton, 1992), p. 253.
4. C. L. Arthur, L. Killam, S. Motlagh, M. Lim, D. Potter and J. Pawliszyn, "Analysis of Substituted Benzene Compounds in Groundwater Using Solid Phase Microextraction" *Environ. Sci. Technol.* **26,** 979 (1992).
5. D. Louch, S. Motlagh and J. Pawliszyn, "Dynamics of Organic Compound Extraction from Water Using Solid Phase Microextraction with Fused Silica Fibers" Anal. Chem. **64,** 1187 (1992).
6. S. Hawthorne, D. Miller, J. Pawliszyn and C. Arthur, "Solventless Determination of Caffeine in Beverages Using Solid Phase Microextraction with Fused Silica Fibres" *J. Chromatogr.* **603,** 185 (1992).
7. C. L. Arthur, L. Killam, K. Buchholz and J. Pawliszyn, "Automation and Optimization of Solid Phase Microextraction" *Anal. Chem.* **64,** 1960 (1992).
8. C. L. Arthur, D. Potter, K. Buchholz, S. Motlagh and J. Pawliszyn, "Solid Phase Microextraction for the Direct Analysis of Water: Theory and Practice" *LC-GC* **10,** 656 (1992).
9. C. L. Arthur, R. Belardi, K. Pratt, S. Motlagh and J. Pawliszyn, "Environmental Analysis of Organic Compounds in Water Using Solid Phase Microextraction" *J. High Resolut. Chromatogr.* **15,** 741 (1992).
10. D. Potter and J. Pawliszyn, "Detection of Substituted Benzenes in Water at the sub pg level Using Solid Phase Microextraction and Gas Chromatography-Ion Trap Mass Spectrometry" *J. Chromatogr.* **625,** 247 (1992).
11. C. L. Arthur, L. Killam, K. Buchholz, D. Potter, M. Chai, Z. Zhang and J. Pawliszyn, "Solid Phase Microextraction: An Attractive Alternative, or Applications of Automated SPME to Environmental Analysis" *Amer. Environ. Lab.* **11,** 10 (1992).
12. E. O. Otu and J. Pawliszyn, "Solid Phase Microextraction of Metal Ions" *Mikrochim. Acta.* **112,** 41 (1993).
13. J. P. Conzen, J. Burck and H. J. Ache, "Characterization of a Fiber-Optic Evanescent Wave Absorbance Sensor for Nonpolar Organic Compounds" *Appl. Spectrosc.* **47,** 753 (1993).

14. Z. Zhang and J. Pawliszyn, "Headspace Solid Phase Microextraction" *Anal. Chem.* **65,** 1843 (1993).
15. C. L. Arthur, M. Chai and J. Pawliszyn, "Solventless Injection Techniques for Microcolumn Separations" *J. Microcolumn Sep.* **5,** 51 (1993).
16. B. D. Page and G. Lacroix, "Application of solid-phase microextraction to the headspace gas chromatographic analysis of halogenated volatiles in selected foods" *J. Chromatogr.* **648,** 199 (1993).
17. J. Berg, "Practical Use of Automated Solid Phase Microextraction" *Amer. Lab.* **11,** 18 (1993).
18. M. Chai, C. Arthur, J. Pawliszyn, R. Belardi and K. Pratt, "Determination of Volatile Chlorinated Hydrocarbons in Air and Water with Solid-Phase Microextraction" *Analyst* **118,** 1501 (1993).
19. K. D. Buchholz and J. Pawliszyn, "Determination of Phenols by Solid Phase Microextraction and Gas Chromatographic Analysis" *Environ. Sci. Technol.* **27,** 2844 (1993).
20. Z. Zhang and J. Pawliszyn, "Analysis of Organic Compounds in Environmental Samples Using Headspace Solid Phase Microextraction" *J. High Resolut. Chromatogr.* **16,** 689 (1993).
21. S. Motlagh and J. Pawliszyn, "On-Line Monitoring of Flowing Samples Using SPME-GC" *Anal. Chim. Acta* **284,** 265 (1993).
22. K. D. Buchholz and J. Pawliszyn, "Optimization of Solid Phase Microextraction Conditions for Determination of Phenols" *Anal. Chem.* **66,** 160 (1994).
23. G. Kraus, A. Brecht, V. Vasic and G. Gaugliz, "Polymer based RIFS sensing: an approach to the indirect measurement of organic pollutants in water" *Fresenius J. Anal. Chem.* **348,** 598 (1994).
24. D. Potter and J. Pawliszyn, "Rapid Determination of Polyaromatic Hydrocarbons and Polychlorinated Biphenyls in Water using solid Phase Microextraction and GC-MS" *Environ. Sci. Technol.* **28,** 298 (1994).
25. B. Schaefer and W. Engewald, "Enrichment of nitroaromatic compounds from water samples by SPME" *GIT Spez. Chromatogr.* **14,** X85 (1994).
26. R. Shirey, V. Mani and J. Savrock, "SPME: A solventless Sample Preparation method for Organic compounds in water" *American Environmental Lab*, 43 (1994).
27. M. E. Cisper, W. L. Earl, N. S. Nogar and P. H. Hemberger, "Silica-Fiber Microextraction for Laser Desorption ITMS" *Anal. Chem.* **66,** 1897 (1994).
28. X. Yang and T. Peppard, "SPME for Flavor Analysis" *J. Agric. Food Chem.* **42,** 1925 (1994).
29. B. MacGillivray, J. Pawliszyn, P. Fowlie and C. Sagara, "Headspace SPME vs Purge and Trap for the Determination of Substituted Benzenes in Water" *J. Chromatogr. Sci.* **32,** 317 (1994).
30. L. P. Sarna, G. R. B. Webster, M. R. Friesen-Fischer and R. S. Ranjan, "Analysis of the petroleum components BTEX in water by commercially available SPME and carbon-layer open tublar capillary column GC" *J. Chromatogr. A* **677,** 201 (1994).

31. Z. Zhang, M. J. Yang and J. Pawliszyn, "Solid Phase Microextraction" *Anal. Chem.* **66,** 844 (1994).
32. J. Horng and S. Huang, "Determination of the semi-volatile compounds nitrobenzene, isophorone, 2,4-dinitrotoluene and 2,6-dinitrotoluene in water using SPME with a PDMS-coated fibre" *J. Chromatogr. A* **678,** 313 (1994).
33. H. B. Wan, H. Chi, M. K. Wong and C. Y. MoK, "SPME using pencil lead as sorbent for analysis of organic pollutants in water" *Anal. Chim. Acta* **298,** 219 (1994).
34. T. Nilsson, F. Pelusio, L. Montanarella, B. Larsen, S. Facchetti and J. O. Madsen, "An evaluation of solid–phase microextraction for analysis of volatile organic compounds in drinking water" *J. High Resolut. Chromatogr.* **18,** 617 (1995).
35. B. L. Wittkamp and D. C. Tilotta, "Determination of BTEX Compounds in Water by SPME and Raman Spectroscopy" *Anal. Chem.* **67,** 600 (1995).
36. Y. Cai and J. M. Bayona, "Determination of methylmercury in fish and river water samples using in situ sodium tetraethylborate derivatization followed by SPME and GC-MS" *J. Chromatogr. A* **696,** 113 (1995).
37. H. M. Yan, K. Kraus and G. Gauglitz, "Detection of Mixtures of Organic Pollutants in water by polymer film receptors in fibre-optical sensors based on reflectometric interference spectrometry" *Anal. Chim. Acta* **312,** 1 (1995).
38. B. Schaefer and W. Engewald, "Enrichment of nitrophenols from water by means of SPME" *Fresenius' J. Anal. Chem.* **352,** 535 (1995).
39. L. Pan, M. A. Adams and J. Pawliszyn, "Determination of Fatty Acids using SPME" *Anal. Chem.* **67,** 4396 (1995).
40. J. R. Dean, W. R. Tomlinson, V. Makovakaya and R. Cumming, "SPME as a Method for Estimating the Octanol-Water Partition Coefficient" *Anal. Chem.* **68,** 130–133 (1996).
41. J. J. Langenfeld, S. B. Hawthorne and D. J. Miller, "Quantitative Analysis of Fuel-Related Hydrocarbons in Surface Water and Wastewater Samples by SPME" *Anal. Chem.* **68,** 144 (1996).
42. K. N. Graham, L. P. Sarna, G. R. B. Webster, J. P. Grynar and H. Y. F. Ng, "SPME of the herbicide metolachlor in runoff and tile-drainage water samples" *J. Chromatogr. A* **725,** 129 (1996).
43. D. C. Heglund and D. C. Tilotta, "Determination of VOCs in Water by SPME and Infrared Spectroscopy" *Environ. Sci. Technol.* **30,** 1212 (1996).
44. A. A. Boyd-Boland and J. Pawliszyn, "SPME coupled with HPLC for the Determination of Alkylphenol Ethoxylate Surfactants in Water" *Anal. Chem.* **68,** 1521 (1996).
45. S. P. Thomas, R. R. Sri, G. R. B. Websterrie and L. S. P., "Protocol for the Analysis of High Concentrations of BTEX in Water using Automated SPME-GC-FID" *Environ. Sci. Technol.* **30,** 1521 (1996).

46. S. S. Johansen and J. Pawliszyn, "Trace Analysis of Hetero Aromatic Compounds in Water by SPME" *J. High Resolut. Chromatogr.* **19,** 627 (1996).
47. T. Gorecki and J. Pawliszyn, "Determination of Tetraethyllead and Inorganic Lead in Water by Solid Phase Microextraction Gas Chromatography" *Anal. Chem.* **68,** 3008 (1996).
48. A. A. Boyd-Boland and J. B. Pawliszyn, "Solid-Phase Microextraction Coupled with High-Performance Liquid Chromatography for the Determination of Alkylphenol Ethoxylate Surfactants in Water" *Anal. Chem.* **68**(9), 1521 (1996).
49. L. Urruty and M. Montury, "Influence of Ethanol on Pesticide Extraction in Aqueous Solutions by Solid Phase Microextraction" *J. Agric. Food Chem.* **44,** 3871 (1996).
50. F. Guo, T. Gorecki, D. Irish and J. Pawliszyn, "Solid-phase microextraction combined with electrochemistry" *Anal. Commun* **33,** 361 (1996).
51. K. J. Hageman, L. Mazeeas, C. B. Grabanski, D. J. Miller and S. B. Hawthorne, "Coupled Subcritical Water Extraction with Solid Phase Microextraction for Determining Semivolatile Organic in Environmental Solids" *Anal. Chem.* **68,** 3892 (1996).
52. Z. Penton, H. Geppet and V. Betz, "Solid Phase Microextraction trace analysis of phenols and pesticides in water with automated SPME and agitation" *Spez. Chromatogr.* **16,** 112 (1996).
53. J. Czerwinski, B. Zygmunt and J. Namiesnik, "Application of solid phase microextraction for determination of volatile halogenated hydrocarbons in air and water of an indoor swimming pool" *Fresenius Environ. Bull.* **5,** 55 (1996).
54. K. Kadokami, K. Sato, T. Iwamura and Y. Hanada, "Determination of hydrophilic alcohols from aquatic environment by solid phase microextraction and GC/MS" *Bunseki Kagaku* **45,** 1013 (1996).
55. R. Eisert and K. Levsen, "Solid-phase microextraction of organic trace amounts from aqueous environmental samples" *GIT Fachz. Lab.* **40,** 581, 586 (1996).
56. T. Gorecki, R. Mindrup and J. Pawliszyn, "Pesticides by solid phase microextraction. Results of a round robin test" *Analyst* **121,** 1381 (1996).
57. W. Vaes, C. Hamwijk, E. U. Ramos, H. J. M. Verhaar and J. L. M. Hermens, "Partitioning of Organic Chemicals to Polyacrylate-Coated Solid Phase Microextraction Fibers: Kinetic Behavior and Quantitative Structure-Property Relationships" *Anal. Chem.* **68,** 4459 (1996).
58. I. Valor, C. Cortada and J. C. Molto, "Direct solid phase microextraction for the determination of BTEX in water and wastewater" *J. High Resolut. Chromatogr.* **19,** 472 (1996).
59. F. J. Santos, M. T. Galceran and D. Fraisse, "Application of solid-phase microextraction to the analysis of volatile organic compounds in water" *J. Chromatogr., A,* **742,** 181, (1996).
60. T. Gorecki, A. Boyd-Boland, Z. Zhang and J. Pawliszyn, "1995 McBryde

Medal Award Lecture. Solid-phase microextraction-a unique tool for chemical measurements" *Can. J. Chem.* **74,** 1297 (1996).
61. T. K. Choudhury, K. O. Gerhardt and T. P. Mawhinney, "Solid-Phase Microextraction of Nitrogen- and Phosphorus-Containing Pesticides from Water and Gas Chromatographic Analysis" *Environ. Sci. Technol.* **30,** 3259 (1996).
62. J. Czerwinski, B. Zygmunt and J. Namiesnik, "Head-space solid phase microextraction for the GC-MS analysis of terpenoids in herb-based formulations" *Fresenius' J. Anal. Chem.,* **356,** 80 (1996).
63. Z. Zhang, J. Poerschmann and J. Pawliszyn, "Direct solid phase microextraction of complex aqueous samples with hollow fiber membrane protection" *Anal. Commun.* **33,** 219 (1996).
64. G. Maurizio, "Use of solid-phase microextraction (SPME) in determination of chlorobenzenes in water samples" *Boll. Chim. Ig., Parte Sci.* **47,** 37 (1996).
65. R. Nakagawa and A. Saito, "A study on occurrence of trihalomethanes in tap waters of Chiba Prefecture [Japan]" *Kankyo Kagaku Kenkyu Hokoku* (Chiba Daigaku) 1995, **21,** 23 (1996).
66. K. Pratt, R. Shirey and V. Mani, "Solid-phase microextraction of VOCs in water" *ASTM Spec. Tech. Publ.,* STP **1261** (Volatile Organic Compounds in the Environment), 139 (1996).
67. J. Ritter, V. K. Stromquist, H. T. Mayfield, M. V. Henley and B. K. Lavine, "Solid phase microextraction for monitoring jet fuel components in groundwater" *Microchem. J.* **54,** 59 (1996).
68. A. A. Boyd-Boland, S. Magdic and J. Pawliszyn, "Simultaneous determination of 60 pesticides in water using solid-phase microextraction and gas chromatography-mass spectrometry" *Analyst* (Cambridge, U. K.) **121,** 929 (1996).
69. R. Eisert, K. Levsen and G. Wuensch, "Multi-residue method for the determination of organic trace pollutants in aqueous samples by solid-phase microextraction and gas chromatography" *Vom Wasser* **86,** 1 (1996).
70. E. Fattore, E. Benfenati and B. Fanelli, "Analysis of chlorinated 1,3-butadienes by solid-phase microextraction and gas chromatography-mass spectrometry" *J. Chromatogr., A* **737,** 85 (1996).
71. R. Eisert and K. Levsen, "Development of a prototype system for quasi-continuous analysis of organic contaminants in surface or sewage water based on in-line coupling of solid-phase microextraction to gas chromatography" *J. Chromatogr., A* **737,** 59 (1996).
72. Y. Hanai, A. Saitou and N. Mio, "Determination of iodine trihalomethanes in drinking water" *Yokohama Kokuritsu Daigaku Kankyo Kagaku Kenkyu Senta Kiyo* **22,** 11 (1996).
73. R. Young, V. Lopez-Avila and W. F. Beckert, "Online determination of organochlorine pesticides in water by solid-phase microextraction and gas chromatography with electron capture detection" *J. High Resolut. Chromatogr.* **19,** 247 (1996).

74. S. Magdic, A. Boyd-Boland, K. Jinno and J. B. Pawliszyn, "Analysis of organophosphorus insecticides from environmental samples using solid-phase microextraction" *J. Chromatogr., A* **736,** 219 (1996).
75. Z. Zhang and J. Pawliszyn, "Sampling volatile organic compounds using a modified solid phase microextraction device" *J. High Resolut. Chromatogr.* **19,** 155 (1996).
76. Z. E. Penton, "Three modes of sample introduction with a single GC autosampler. Liquid injection, headspace, and solid-phase microextraction" *Am. Lab.* (Shelton, Conn.) **28,** 18B (1996).
77. R. Eisert and K. Levsen, "Solid-phase microextraction coupled to gas chromatography: a new method for the analysis of organics in water" *J. Chromatogr., A* **733,** 143 (1996).
78. H. Naito, S. Kadowaki and R. Suzuki, "GC/MS analysis of polycyclic aromatic hydrocarbons in water sample using solid phase microextraction" *Aichi-ken Kogai Chosa Senta Shoho* **23,** 49 (1995).
80. K. N. Graham, L. P. Sarna, G. R. B. Webster, J. D. Gaynor and H. Y. F. Ng, "Solid-phase microextraction of the herbicide metolachlor in runoff and tile-drainage water samples" *J. Chromatogr., A* **725,** 129 (1996).
81. K. K. Chee, M. K. Wong and H. K. Lee, "Determination of 4-nonylphenol—Part 2: orthogonal array design as a chemometric method for the solid-phase microextraction of 4-nonylphenol in water" *J. Microcolumn Sep.* **8,** 131 (1996).
82. S. Magdic, A. A. Boyd-Boland and J. B. Pawliszyn, "The analysis of pesticides using solid phase microextraction" *Organohalogen Compd.* **23,** 47 (1995).
83. S. Tutschku, S. Mothes and R. Wennrich, "Preconcentration and determination of Sn- and Pb-organic species in environmental samples by SPME and GC-AED" *Fresenius' J. Anal. Chem.* **354,** 587 (1996).
84. Y. Morcillo, Y. Cai and J. M. Bayona, "Rapid determination of methyltin compounds in aqueous samples using solid phase microextraction and capillary gas chromatography following in-situ derivatization with sodium tetraethylborate" *J. High Resolut. Chromatogr.* **18**(12), 767 (1995).
85. S. P. Thomas, R. Sri Ranjan, G. R. B. Webster and L. P. Sarna, "Protocol for the Analysis of High Concentrations of Benzene, Toluene, Ethylbenzene, and Xylene Isomers in Water Using Automated Solid-Phase Microextraction-GC-FID" *Environ. Sci. Technol.* **30,** 1521 (1996).
86. W. Bechmann and P. Volkmer, "The application of solid-phase microextraction (SPME) for the determination of triazines" *GIT Fachz. Lab* **39,** 1129, 1132 (1995).
87. P. Popp, S. Mothes and L. Brueggemann, "Determination of pesticides in water by solid-phase microextraction and gas chromatography" *Vom Wasser* **85,** 229 (1995).
88. D. L. Heglund and D. C. Tilotta, "Determination of Volatile Organic Compounds in Water by Solid Phase Microextraction and Infrared Spectroscopy" *Environ. Sci. Technol.* **30**(4), 1212 (1996).

89. P. Popp, A. Paschke, U. Schroeter and G. Oppermann, "Application of solid-phase microextraction (SPME) in combination with GC/FID to the determination of benzene and halogenated benzenes in pure and octanol-saturated water" *Chem. Anal.* (Warsaw) **40,** 897, (1995).
90. K. Levsen, R. Eisert, S. Sennert and D. Volmer, "Analytical methods for the determination of pesticides and their occurrence in the ground water" *Proc. SPIE-Int. Soc. Opt. Eng.* **2504** (Environmental Monitoring and Hazardous Waste Site Remediation, 1995), 127 (1995).
91. R. E. Shirey, "Rapid analysis of environmental samples using solid-phase microextraction (SPME) and narrow bore capillary columns" *J. High Resolut. Chromatogr.* **18,** 495 (1995).

Food, Natural Products, and Pharmaceutical Analysis

1. S. Hawthorne, D. Miller, J. Pawliszyn and C. Arthur, "Solventless Determination of Caffeine in Beverages Using Solid Phase Microextraction with Fused Silica Fibres" *J. Chromatogr.* **603,** 185 (1992).
2. B. D. Page and G. Lacroix, "Application of solid-phase microextraction to the headspace gas chromatographic analysis of halogenated volatiles in selected foods" *J. Chromatogr.* **648,** 199 (1993).
3. X. Yang and T. Peppard, "SPME for Flavor Analysis" *J. Agric. Food Chem.* **42,** 1925 (1994).
4. R. Mindrup, "Measure Flavors Using SPME" *Food Qual.,* 40 (1995).
5. V. Mani and C. Wooley, "SPME and chiral separations for food and flavors" *Foods Food Ingredients J. Jpn* **163,** 94 (1995).
6. R. F. Mindrup, "SPME simplifies preparation of forensic, pharmaceutical, and food and beverage samples" *Chem. N. Z.* **59,** 21 (1995).
7. R. E. Shirey, C. L. Woolley and R. F. Mindrup, "Solventless Extraction of Flavors and Other Food Components" *Food Test. Anal.* 39 (1995).
8. X. Yang and T. Peppard, "SPME of Flavor Compounds—A comparison of two fiber coatings and a discussion of the rules of thumb for adsorption" *LC-GC* **13,** 882 (1995).
9. Z. Penton, "Flavor volatiles in a fruit beverage with automated SPME" *Food Test. Anal.* **2,** 16 (1996).
10. W. Vaes, H. J. Wouter, E. U. Ramos, H. J. M. Verhaar, W. Seinen and J. L. M. Hermens, "Measurement of the Free Concentration Using Solid Phase Microextraction: Binding to Protein" *Anal. Chem.* **68,** 4463 (1996).
11. V. Mani, R. Shirey, C. Woolley, C. Jabco, W. Ramsey and G. Wachob, "New application of SPME/HPLC" *Lab 2000* **10,** 66 (1996).
12. A. J. Matich, D. D. Rowan and N. H. Banks, "Solid Phase Microextraction for Quantitative Headspace Sampling of Apple Volatiles" *Anal. Chem.* **68,** 4114 (1996).
13. E. Ibanez and R. A. Bernhard, "Solid-phase microextraction (SPME) of pyrazines in model reaction systems" *J. Sci. Food Agric.* **72,** 91 (1996).

14. A. K. Karlson-Borg and R. Mozuraitis, "Solid phase microextraction technique used for collecting semiochemicals. Identification of volatiles released by individual signaling Phyllonorycter sylvella moths" *Z. Naturforsch., C: Biosci.* **51,** 599 (1996).
15. T. Nilsson, T. O. Larsen, L. Montanarella and J. O. Madsen, "Application of Head-space solid-phase microextraction for the analysis of volatile metabolites emitted by Penicillium species" *J. Microbiol. Methods* **25,** 245 (1996).
16. K. G. Miller, C. F. Poole and T. M. P. Pawlowski, "Classification of the botanical origin of cinnamon by solid-phase microextraction and gas chromatography" *Chromatographia* **42,** 639 (1996).
17. A. Steffen and J. Pawliszyn, "Analysis of Flavor Volatiles Using Headspace Solid-Phase Microextraction" *J. Agric. Food Chem.* **44,** 2187 (1996).
18. D. De Ia Calle Garcia, S. Magnaghi, M. Reichenbaecher and K. Danzer, "Systematic optimization of the analysis of wine bouquet components by solid-phase microextraction" *J. High Resolut. Chromatogr.* **19,** 257 (1996).
19. J. A. Field, G. Nickerson, D. D. James and C. Heider, "Determination of essential oils in hops by headspace solid-phase microextraction" *J. Agric. Food Chem.* **44,** 1768 (1996).
20. W. M. Coleman, III, "A study of the behavior of Maillard reaction products analyzed by solid-phase microextraction-gas chromatography-mass selective detection" *J. Chromatogr. Sci* **34,** 213 (1996).
21. L. K. Ng, M. Hupe, J. Harnois and D. Moccia, "Characterization of commercial vodkas by solid-phase microextraction and gas chromatography/mass spectrometry analysis" *J. Sci. Food Agric.* **70,** 380 (1996).
22. N. Yasuda, K. Otsuki, M. Nishikawa, M. Katagi and H. Tsuchihashi, "Analysis of components in crude drugs by headspace solid phase micro extraction method. I" *Yakugaku Zasshi* **116,** 251 (1996).
23. D. Ulrich, S. Eunert, E. Hoberg and A. Rapp, "Analysis of strawberry aroma by solid phase micro extraction" *Dtsch. Lebensm.-Rundsch.* **91,** 349 (1995).
24. H. Hisano, K. Ishimaru, H. Tada and Y. Ikeda, "Flavors of heliotrope flowers, analyzed by solid phase microextraction method" *Nippon Shokuhin Kagaku Gakkaishi* **2,** 6 (1995).
25. D. Picque, A. Normand and G. Corrieu, "Evaluation of the solid phase microextraction for the direct analysis of aroma compounds in banana" *Colloq.-Inst. Natl. Rech. Agron.* **75** (Bioflavour 95) 117 (1995).

Clinical and Forensic Analysis

1. M. Yashiki, T. Miyazaki and T. Kojima, "Detection of amphetamines and inflammable compounds in biological materials using GC/MS and SPME" *Hochudoku* **12,** 120 (1994).

2. M. Chiaroti and R. Marsili, "GC analysis of methadone in urine samples after SPME (also in Proc. 16th Int. Symp. on Cap. Chrom., 191994, pages 892–896)" *J. Microcolumn Sep.* **6,** 577 (1994).
3. X. P. Lee, T. Kumazawa and K. Sato, "A Simple Analysis of 5 Thinner Components in Human-Body Fluids by Headspace Solid Phase Microextraction (SPME)" *Int. J. of Legal Medicine* **107,** 310 (1995).
4. R. F. Mindrup, "SPME simplifies preparation of forensic, pharmaceutical, and food and beverage samples (also published in Supelco Reporter 14(1) 1995)" *Chem. N. Z.* **59,** 21 (1995).
5. T. Kumazawa, X. Lee, K. Sato, H. Seno, A. Ishii and O. Suzuki, "Detection of ten local anesthetics in human blood using solid-phase microextraction (SPME) and capillary gas chromatography" *Jpn. J. Forensic Toxicol.* **13,** 182 (1995).
6. M. Yashiki, N. Nagasawa, T. Kojima, T. Miyazaki and Y. Iwasaki, "Rapid Analysis of nicotine and continine in urine using headspace SPME and selected ion monitoring" *Jpn. J. Forensic Toxicol.* **13,** 17 (1995).
7. T. Kumazawa, K. Watanabe, K. Sato, H. Seno, A. Ishii and O. Suzukil, "Detection of cocaine in human urine by Solid-phase microextraction and capillary gas chromatography with nitrogen-phosphorus detection" *Jpn. J. Forensic Toxicol.* **13,** 207 (1995).
9. M. Yashiki, T. Kojima, T. Miyazaki, N. Nagasawa, Y. Iwasaki and K. Hara, "Detection of amphetamines in urine using head space-solid phase microextraction and chemical ionization selected ion monitoring" *Forensic Sci. Int.* **76,** 169 (1995).
10. M. Krogh, K. Johansen, F. Tonnesen and K. E. Rasmussen, "Solid-phase microextraction for the determination of the free concentration of valproic acid in human plasma by capillary gas chromatography" *J. Chromatogr., B: Biomed. Appl.* **673,** 299 (1995).
11. N. Nagasawa, M. Yashiki, Y. Iwasaki, K. Hara and T. Kojima, "Rapid analysis of amphetamines in blood using solid phase microextraction (SPME)" *Nippon Iyo Masu Supekutoru Gakkai Koenshu* **20,** 127 (1995).
12. T. Kaneko and M. Nakada, "Forensic application of the solid-phase microextraction method to the analysis of gasoline and kerosine" *Kagaku Keisatsu Kenkyusho Hokoku, Hokagaku-hen* **48,** 107 (1995).
13. H. Seno, T. Kumazawa, A. Ishii, M. Nishikawa, H. Hattori and O. Suzuki, "Detection of meperidine (pethidine) in human blood and urine by headspace solid phase microextraction and gas chromatography" *Jpn. J. Forensic Toxicol.* **13,** 211 (1995).
14. T. Kumazawa, K. Watanabe, K. Sato, H. Seno, A. Ishii and O. Suzuki, "Detection of cocaine in human urine by solid-phase microextraction and capillary gas chromatography with nitrogen-phosphorus detection" *Jpn. J. Forensic Toxicol.* **13,** 207 (1995).
15. K. G. Furton, J. Bruna and J. R. Almirall, "A simple, inexpensive, rapid, sensitive and solventless technique for the analysis of accelerants

in fire debris based on SPME" *J. High Resolut. Chromatogr.* **18,** 625 (1995).

16. Y. Iwasaki, M. Yashiki, N. Nagasawa, T. Miyazaki and T. Kojima, "Analysis of inflammable substances in blood using head space-solid phase microextraction and chemical ionization selected ion monitoring" *Jpn. J. Forensic Toxicol.* **13,** 189 (1995).
17. D. C. Robacker and R. J. Bartelt, "Solid Phase Microextraction Analysis of Static-Air Emissions of Ammonia, Methylamine, and Putrescine from a Lure for the Mexican Fruit Fly (Anastrepha ludens)" *J. Agric. Food Chem.* **44,** 3554 (1996).
18. K. Ameno, C. Fuke, S. Ameno, H. Kinoshita and I. Ijiri, "Application of a solid phase microextraction technique for the detection of urinary methamphetamine and amphetamine by gas chromatograph" *J.-Can. Soc. Forensic Sci.* **29,** 43 (1996).
19. T. Kumazawa, K. Sato, H. Seno, A. Ishii and O. Suzuki, "Extraction of local anesthetics from human blood by direct immersion-solid phase microextraction (SPME)" *Chromatographia* **43,** 59 (1996).
20. K.G. Furton, J. R. Almirall and J. C. Bruna, "A novel method for the analysis of gasoline from fire debris using headspace solid-phase microextraction" *J. Forensic Sci.,* **41,** 12 (1996).
21. A. Steffen and J. Pawliszyn, "Determination of liquid accelerants in arson suspected fire debris using headspace solid-phase microextraction" *Anal. Commun.* **33,** 129 (1996).
22. H. Seno, T. Kumazawa, A. Ishii, M. Nishikawa, K. Watanabe, H. Hattori and O. Suzuki, "Detection of some phenothiazines by headspace solid phase microextraction and gas chromatography" *Jpn. J. Forensic Toxicol.* **14,** 30 (1996).
23. N. Xu, S. Vandegrift and G. W. Sewell, "Determination of chloroethenes in environmental biological samples using gas chromatography coupled with solid phase micro extraction" *Chromatographia* **42,** 313 (1996).
24. X. P. Lee, T. Kumazawa, K. Sato and O. Suzuki, "Detection of organophosphate pesticides in human body fluids by headspace solid-phase microextraction (SPME) and capillary gas chromatography with nitrogen-phosphorus detection" *Chromatographia* **42,** 135 (1996).
25. J. L. Liao, C. M. Zeng, S. Hjerten and J. Pawliszyn, "Solid phase microextraction of biopolymers, exemplified with adsorption of basic proteins onto a fiber coated with polyacrylic acid" *J. Microcolumn Sep.* **8,** 1 (1996).
26. I. Maier, G. Pohnert, S. Pantke-Boecker and W. Boland, "Solid-phase microextraction and determination of the absolute configuration of the Laminaria digitata (Laminariales, Phaeophyceae) spermatozoid-releasing pheromone" *Naturwissenschaften* **83,** 378 (1996).
27. B. Schaefer, P. Hennig and W. Engewald, "Analysis of monoterpenes from conifer needles using solid phase microextraction" *J. High Resolut. Chromatogr.* **18,** 587 (1995).

General SPME Theory and Applications

1. R. Vanderhaghen, S. Cueille, B. Drevillon and R. Ossikovski, "Modulate photoellipsometry. Application to the measurement of GaAs internal field" *Phys. Status Solidi A* **152,** 85 (1995).
2. F. Mangani and R. Cenciarini, "Solid phase microextraction using fused silica fibers coated with graphitized carbon black" *Chromatographia* **41,** 678 (1995).
3. P. Popp and G. Oppermann, "Solid-phase microextraction (SPME) connected with GC" *CLB Chem. Labor Biotech.* **47,** 358 (1996).
4. J. J. Langenfeld, S. B. Hawthorne and D. J. Miller, "Optimizing split/splitless injection port parameters for solid-phase microextraction" *J. Chromatogr., A* 740(1), 139–145 (1996).
5. E. D. Conte, D. W. Miller and W. Dwight, "A solid phase microextraction-electrodeposition device for the determination of putrescine and cadaverine by high-resolution gas chromatography" *J. High Resolut. Chromatogr.* **19,** 294 (1996).
6. H. Anderegg, "Introduction into the headspace gas chromatography. Part 6" *Schweiz. Lab.-Z.* **53,** 55 (1996).
7. Z. Zhang and J. Pawliszyn, "Studying Activity Coefficients of Probe Solutes in Selected Liquid Polymer Coatings Using Solid Phase Microextraction" *J. Phys. Chem.* **100,** 17648 (1996).

Glossary

a	Fibre coating inner radius
A	Area of needle opening
A_f	Fiber coating outer surface area
b	Fiber coating outer radius
BTEX	Abbreviation for benzene, toluene, ethylbenzene, and three xylene isomers: *m*-xylene, *o*-xylene, and *p*-xylene
C_f^∞	Equilibrium concentration of analyte in fiber coating
C_g^∞	Equilibrium concentration of analyte in gas
C_h^∞	Equilibrium concentration of analyte in gaseous headspace above sample
C_s^∞	Equilibrium concentration of analyte in sample
C_w^∞	Equilibrium concentration of analyte in water
C_0	Initial concentration of analyte in sample
CE	Capillary electrophoresis
d	Vial inner radius
D_e.	Eddy diffusion coefficient of analyte in sample matrix
D_f.	Diffusion coefficient of analyte in fiber coating
D_g	Diffusion coefficient of analyte in gas
D_s	Diffusion coefficient of analytes in sample matrix
EPA	Environmental Protection Agency of the United States of America
FID	Flame ionization detector, commonly used in a gas chromatograph
GC	Gas chromatography, or gas chromatograph
HPLC	High performance liquid chromatography
i.d.	Inside diameter

ITMS	Ion trap mass spectrometer
k	Pseudo first order rate constant of chemical reaction
k'	Rate constant of chemical reaction
k_p	Partition ratio ($k_p = K_{fs}V_f/V_V$)
K_{fg}	Fiber/gas distribution constant ($K_{fg} = C_f^\infty/C_g^\infty$)
K_{fh}	Fiber/headspace distribution constant ($K_{fh} = C_f^\infty/C_h^\infty$)
K_{fs}	Fiber/sample matrix distribution constant ($K_{fs} = C_f^\infty/C_s^\infty$)
K_{fw}	Fiber/water distribution constant ($K_{fs} = C_f^\infty/C_w^\infty$)
K_{hs}	Headspace/sample distribution constant ($K_{hs} = C_h^\infty/C_s^\infty$)
K_F	Henry's law constant of analyte in fiber coating
K_H	Henry's law constant of analyte in sample matrix
L	Fibre coating length
LOD	Limit of detection
MESI	Membrane Extraction with a Sorbent Interface
MS	Mass spectrometry or mass spectrometer
MS/MS	Tandem mass spectrometry
n	Amount of analyte extracted onto the coating
o.d.	Outside diameter
PA	Poly(acrylate)
PAH	Polynuclear aromatic hydrocarbon
PDMS	Poly(dimethylsiloxane)
PDAM	Pyrenyldiazomethane
PTV	Programmable temperature vaporizer (GC injector)
RSD	Relative standard deviation
SD	Standard deviation
SFC	Supercritical fluid chromatography
SFE	Supercritical fluid extraction
S/N	Signal to noise ratio
SPE	Solid phase extraction
SPI	Septum-equipped temperature programmable injector, used in the Varian GC
SPME	Solid phase microextraction
SS	Stainless steel
t_e	Equilibration time
$t_{95\%}$	Time required to extract 95% of analyte amount compared to at equilibrium conditions
u	Velocity of sample matrix in respect to extracting phase
V_f	Volume of fiber coating
V_h	Volume of gaseous headspace above sample
V_s	Sample volume
V_V	Void volume of the tubing containing extracting phase
	VOC Volatile organic compounds
Z	Distance between needle opening and position of coating
δ	Boundary layer thickness

Index